SpringerBriefs in Statistics

JSS Research Series in Statistics

Editors-in-Chief

Naoto Kunitomo
Akimichi Takemura

Series editors

Genshiro Kitagawa
Tomoyuki Higuchi
Yutaka Kano
Toshimitsu Hamasaki
Shigeyuki Matsui
Manabu Iwasaki
Yasuhiro Omori
Yoshihiro Yajima

The current research of statistics in Japan has expanded in several directions in line with recent trends in academic activities in the area of statistics and statistical sciences over the globe. The core of these research activities in statistics in Japan has been the Japan Statistical Society (JSS). This society, the oldest and largest academic organization for statistics in Japan, was founded in 1931 by a handful of pioneer statisticians and economists and now has a history of about 80 years. Many distinguished scholars have been members, including the influential statistician Hirotugu Akaike, who was a past president of JSS, and the notable mathematician Kiyosi Itô, who was an earlier member of the Institute of Statistical Mathematics (ISM), which has been a closely related organization since the establishment of ISM. The society has two academic journals: the Journal of the Japan Statistical Society (English Series) and the Journal of the Japan Statistical Society (Japanese Series). The membership of JSS consists of researchers, teachers, and professional statisticians in many different fields including mathematics, statistics, engineering, medical sciences, government statistics, economics, business, psychology, education, and many other natural, biological, and social sciences.

The JSS Series of Statistics aims to publish recent results of current research activities in the areas of statistics and statistical sciences in Japan that otherwise would not be available in English; they are complementary to the two JSS academic journals, both English and Japanese. Because the scope of a research paper in academic journals inevitably has become narrowly focused and condensed in recent years, this series is intended to fill the gap between academic research activities and the form of a single academic paper.

The series will be of great interest to a wide audience of researchers, teachers, professional statisticians, and graduate students in many countries who are interested in statistics and statistical sciences, in statistical theory, and in various areas of statistical applications.

More information about this series at http://www.springer.com/series/13497

Masafumi Akahira

Statistical Estimation for Truncated Exponential Families

Springer

Masafumi Akahira
Institute of Mathematics
University of Tsukuba
Tsukuba, Ibaraki
Japan

ISSN 2191-544X ISSN 2191-5458 (electronic)
SpringerBriefs in Statistics
ISSN 2364-0057 ISSN 2364-0065 (electronic)
JSS Research Series in Statistics
ISBN 978-981-10-5295-8 ISBN 978-981-10-5296-5 (eBook)
DOI 10.1007/978-981-10-5296-5

Library of Congress Control Number: 2017945227

Printed on acid-free paper

This Springer imprint is published by Springer Nature
The registered company is Springer Nature Singapore Pte Ltd.
The registered company address is: 152 Beach Road, #21-01/04 Gateway East, Singapore 189721, Singapore

Preface

In the theory of statistical estimation, the asymptotic properties such as the consistency, asymptotic normality, and asymptotic efficiency have been discussed under usual regularity conditions, and in particular, it is well known that the maximum likelihood estimator (MLE) has such properties. But, in the non-regular case when the regularity conditions do not necessarily hold, it is seen that the property of estimator depends on the irregularity. In the book, we treat truncated exponential families of distributions as a typical situation when both of the regular and non-regular structures exist and clarify how they affect the estimation of a natural parameter and truncation parameters. Such families include a (upper-truncated) Pareto distribution which is widely used in various fields such as finance, physics, hydrology, astronomy, and other disciplines. For a one-sided truncated exponential family (oTEF) with a natural parameter and a truncation parameter, we consider the estimation of a natural parameter with a truncation parameter as nuisance one. Then, the MLE of a natural parameter when a truncation parameter is known and the MLE of a natural parameter when a truncation parameter is unknown have been seen to have the same asymptotic normality. However, in the book it is shown that the asymptotic difference between them appears in the second order after a bias-adjustment, and it is defined as the notion of second-order asymptotic loss through the second-order asymptotic variances obtained from their stochastic expansions. The regular and non-regular structures of the oTEF are reflected in the second-order asymptotic variance of the latter MLE, which effects the loss. The corresponding results to the case of a oTEF are obtained in that of a two-sided truncated exponential family (tTEF) of distributions with a natural parameter and truncation parameters. We also conversely consider the estimation of a truncation parameter with a natural parameter as nuisance one for a oTEF using a bias-adjustment. The bias-adjusted MLE of a truncation parameter when a natural parameter is known and the bias-adjusted MLE of a truncation parameter when a natural parameter is unknown are constructed, and their asymptotic difference is clarified in a similar way to the case of the estimation of a natural parameter. The corresponding results to the case of a oTEF are obtained in the case of a tTEF. From the Bayesian viewpoint, such estimation is also discussed.

Further several examples including a truncated exponential, a truncated normal, a (upper-truncated) Pareto, a truncated beta, and a truncated Erlang (type) distributions are given. In some examples, related results to the uniformly minimum variance unbiased estimation are also described.

Tsukuba, Japan
April 2017

Masafumi Akahira

Acknowledgements

I am grateful to Dr. H. Tanaka and Dr. K. Koike for their comments to the earlier draft and Dr. S. Hashimoto for making the tables and figures with numerical calculation. I would also like to express my thanks to him, Dr. K. Yata and Dr. N. Ohyauchi for their assistance in preparation of the manuscript.

Contents

Chapter 1
Asymptotic Estimation for Truncated Exponential Families

Multiparameter models are considered, truncated exponential families of distributions are defined, and basic notions to compare asymptotically models through estimators are introduced.

1.1 Models with Nuisance Parameters and Their Differences

In parametric models with nuisance parameters, the asymptotic estimation of the parameter of interest was discussed in the regular case (i.e., under the case when the suitable regularity conditions hold) (see, e.g., Barndorff-Nielsen and Cox 1994; Lehmann 1999). The asymptotic theory in regular parametric models with nuisance parameters was discussed by Barndorff-Nielsen and Cox (1994). In higher order asymptotics, under suitable regularity conditions, the concept of asymptotic deficiency discussed by Hodges and Lehmann (1970) is useful in comparing asymptotically efficient estimators in the presence of nuisance parameters. Indeed, the asymptotic deficiencies of some asymptotically efficient estimators relative to the maximum likelihood estimator (MLE) based on the pooled sample were obtained in the presence of nuisance parameters (see, e.g., Akahira and Takeuchi 1982 and Akahira 1986). In the discussion, the notion of orthogonality took an important role (see also Cox and Reid 1987). In order to discriminate asymptotically efficient estimators, the concept of asymptotic deficiency was used as follows. For two estimators $\hat{\theta}_n^{(1)}$ and $\hat{\theta}_n^{(2)}$ of a parameter θ based on a sample of size n, let d_n be an additional size of sample needed such that $\hat{\theta}_n^{(2)}$ is asymptotically equivalent to $\hat{\theta}_n^{(1)}$ in some sense. If $\lim_{n\to\infty} d_n$ exists, it is called the asymptotic deficiency of $\hat{\theta}_n^{(2)}$ relative to $\hat{\theta}_n^{(1)}$, which is useful in comparing asymptotically efficient estimators up to the higher order and investigated by Akahira (1981, 1986, 1992) from the viewpoint of the equivalence of the asymptotic distributions of estimators up to the higher order, under suitable regularity conditions.

M. Akahira, *Statistical Estimation for Truncated Exponential Families*,
JSS Research Series in Statistics, DOI 10.1007/978-981-10-5296-5_1

For example, the asymptotic deficiency is shown to be closely related to the difference between the second-order asymptotic variances of estimators.

In the monograph, we consider a parametric model in which regular and non-regular structures are mixed. Let $X_1, X_2, \ldots, X_n, \ldots$ be a sequence of independently and identically distributed (i.i.d.) random variables with a density $f(x;\theta)$, where $x \in \mathbb{R}^1$, $\theta = (\theta_1, \ldots, \theta_k)$ is a vector-valued parameter in $\mathbb{R}^k$. Now, let θ_1 be a parameter to be interested and $\vartheta := (\theta_2, \ldots, \theta_k)$ is a vector-valued nuisance parameter in $\mathbb{R}^{k-1}$. Here, θ_1 and ϑ are considered as parameters representing the regular and non-regular structures, respectively. Then, we consider two models. When ϑ is known, i.e., $\vartheta = \vartheta_0 := (\theta_{20}, \ldots, \theta_{k0})$, we denote the model by $M_0 := M(\theta_1; \vartheta_0)$. When ϑ is unknown, we denote the model by $M := M(\theta_1; \vartheta)$. So, we consider such an appropriate way to estimate θ_1 as a maximum likelihood method, under the models M_0 and M, and denote it by $\hat{\theta}_{10}$ and $\hat{\theta}_{11}$, respectively. Note that $\hat{\theta}_{11}$ depends on the estimator of ϑ. Suppose that the maximum order of consistency of $\hat{\theta}_{10}$ and $\hat{\theta}_{11}$ is $\sqrt{n}$ and, for each $i = 2, \ldots, k$, that of the estimator $\hat{\theta}_i$ of θ_i is n (see Akahira 1975, 1995, Akahira and Takeuchi 1981, 1995). Let $T_{10} := \sqrt{n}(\hat{\theta}_{10} - \theta_1)$ and $T_{11} := \sqrt{n}(\hat{\theta}_{11} - \theta_1)$. Suppose that the asymptotic means and variances of T_{10} under M_0 and T_{11} under M have expansions of the following type:

$$E_{\theta_1,\vartheta_0}(T_{10}) = \frac{b_{10}(\theta_1, \vartheta_0)}{\sqrt{n}} + O\left(\frac{1}{n\sqrt{n}}\right), \tag{1.1}$$

$$E_{\theta_1,\vartheta}(T_{11}) = \frac{b_{11}(\theta_1, \vartheta)}{\sqrt{n}} + O\left(\frac{1}{n\sqrt{n}}\right), \tag{1.2}$$

$$V_{\theta_1,\vartheta_0}(T_{10}) = v(\theta_1, \vartheta_0) + \frac{c_{10}(\theta_1, \vartheta_0)}{n} + O\left(\frac{1}{n\sqrt{n}}\right), \tag{1.3}$$

$$V_{\theta_1,\vartheta}(T_{11}) = v(\theta_1, \vartheta) + \frac{c_{11}(\theta_1, \vartheta)}{n} + O\left(\frac{1}{n\sqrt{n}}\right) \tag{1.4}$$

respectively. Here and henceforth, the asymptotic mean and asymptotic variance are based on the definition of asymptotic expectation (see Akahira and Takeuchi 1987). Suppose that $b_{10}(\theta_1, \vartheta) = b_{10}(\theta_1, \theta_2, \ldots, \theta_k)$ and $b_{11}(\theta_1, \vartheta) = b_{11}(\theta_1, \theta_2, \ldots, \theta_k)$ are twice continuously differentiable in $(\theta_1, \theta_2, \ldots, \theta_k)$. Let $\hat{\theta}^*_{11}$ be a bias-adjusted estimator of θ_1 such that

$$\hat{\theta}^*_{11} = \hat{\theta}_{11} - \frac{1}{n}\left\{b_{11}(\hat{\theta}_{11}, \hat{\vartheta}) - b_{10}(\hat{\theta}_{11}, \hat{\vartheta})\right\},$$

where $\hat{\vartheta} = (\hat{\theta}_2, \ldots, \hat{\theta}_k)$. Putting $T^*_{11} = \sqrt{n}(\hat{\theta}^*_{11} - \theta_1)$, we have from (1.1), (1.2), and (1.4)

$$E_{\theta_1,\vartheta}(T_{11}^*) = \frac{b_{10}(\theta_1, \vartheta)}{\sqrt{n}} + O\left(\frac{1}{n\sqrt{n}}\right),$$
$$V_{\theta_1,\vartheta}(T_{11}^*) = v(\theta_1, \vartheta) + \frac{c_{11}^*(\theta_1, \vartheta)}{n} + O\left(\frac{1}{n\sqrt{n}}\right),$$

hence, it follows from (1.2) that the asymptotic bias of T_{11}^* is equal to that of T_{10} in order $n^{-1/2}$, where

$$c_{11}^*(\theta_1, \vartheta) = c_{11}(\theta_1, \vartheta) - 2v(\theta_1, \vartheta)\{(\partial/\partial\theta_1)b_{11}(\theta_1, \vartheta) - (\partial/\partial\theta_1)b_{10}(\theta_1, \vartheta)\}.$$

Then, the second-order asymptotic loss of $\hat{\theta}_{11}^*$ relative to $\hat{\theta}_{10}$ is defined by

$$\begin{aligned} d_n(\hat{\theta}_{11}^*, \hat{\theta}_{10}) :=& \frac{n}{v(\theta_1, \vartheta_0)}\left\{V_{\theta_1,\vartheta_0}(T_{11}^*) - V_{\theta_1,\vartheta_0}(T_{10})\right\} \\ =& \frac{c_{11}^*(\theta_1, \vartheta_0) - c_{10}(\theta_1, \vartheta_0)}{v(\theta_1, \vartheta_0)} + o(1) \end{aligned} \tag{1.5}$$

as $n \to \infty$, which is derived from (1.3) and also corresponds to the asymptotic deficiency. The ratio of the asymptotic variance of T_{11}^* to that of T_{10} up to the order $o(1/n)$ is given by

$$R_n(\hat{\theta}_{11}^*, \hat{\theta}_{10}) := \frac{V_{\theta_1,\vartheta_0}(T_{11}^*)}{V_{\theta_1,\vartheta_0}(T_{10})} = 1 + \frac{c_{11}^*(\theta_1, \vartheta_0) - c_{10}(\theta_1, \vartheta_0)}{nv(\theta_1, \vartheta_0)} + o\left(\frac{1}{n}\right) \tag{1.6}$$

as $n \to \infty$. It is easily seen from (1.5) and (1.6) that

$$R_n(\hat{\theta}_{11}^*, \hat{\theta}_{10}) = 1 + \frac{1}{n}d_n(\hat{\theta}_{11}^*, \hat{\theta}_{10}) + o\left(\frac{1}{n}\right)$$

as $n \to \infty$.

Here, we consider the models $M(\theta_1, \vartheta_0)$ and $M(\theta_1, \vartheta)$ and their asymptotic models $M(\hat{\theta}_{10}, \vartheta_0)$ and $M(\hat{\theta}_{11}^*, \vartheta)$. The asymptotic model $M(\hat{\theta}_{10}, \vartheta_0)$ is directly obtained from the model $M(\theta_1, \vartheta_0)$, but the asymptotic model $M(\hat{\theta}_{11}^*, \vartheta_0)$ is given via the asymptotic model $M(\hat{\theta}_{11}^*, \vartheta)$ of $M(\theta_1, \vartheta)$. Then, the difference between the asymptotic models $M(\hat{\theta}_{10}, \vartheta_0)$ and $M(\hat{\theta}_{11}^*, \vartheta_0)$ is represented by (1.5) and (1.6) up to the second order, through the estimator (see Fig. 1.1, Chaps. 2 and 3).

In a similar way to the above, we can conversely consider θ_2 in ϑ as a parameter to be interested and the others in ϑ and θ_1 as nuisance parameters (see Chaps. 4–6).

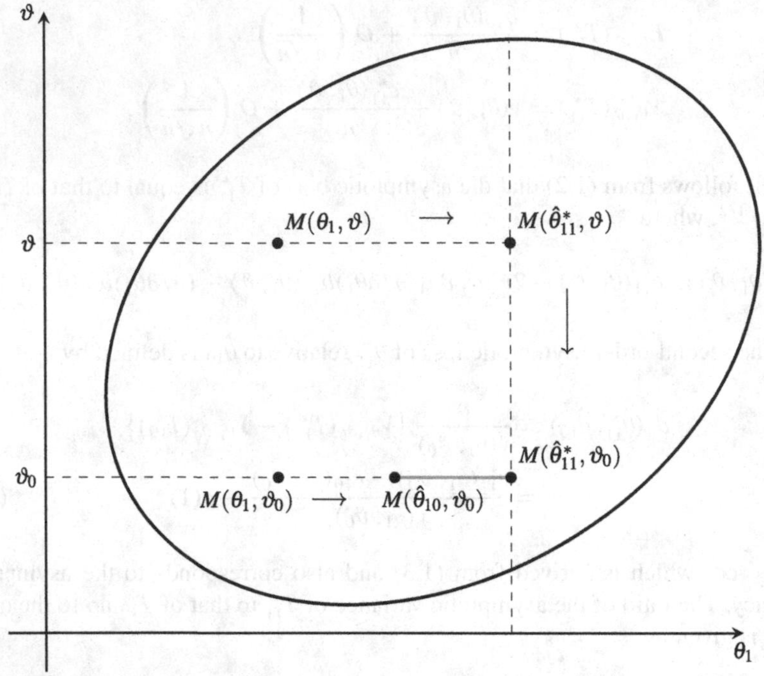

Fig. 1.1 Difference between asymptotic models $M(\hat{\theta}_{10}, \vartheta_0)$ and $M(\hat{\theta}^*_{11}, \vartheta_0)$

1.2 One-Sided Truncated Exponential Family

In a similar way to Bar-Lev (1984), we consider a distribution $P_{\theta,\gamma}$, having density

$$f(x; \theta, \gamma) = \begin{cases} \frac{a(x)e^{\theta u(x)}}{b(\theta,\gamma)} & \text{for } c < \gamma \leq x < d, \\ 0 & \text{otherwise} \end{cases} \tag{1.7}$$

with respect to the Lebesgue measure, where $-\infty \leq c < d \leq \infty$, $a(\cdot)$ is a nonnegative-valued and continuous almost surely, and $u(\cdot)$ is absolutely continuous with $du(x)/dx \not\equiv 0$ over the interval (γ, d). Let

$$\Theta(\gamma) := \left\{ \theta \,\middle|\, 0 < b(\theta, \gamma) := \int_\gamma^d a(x)e^{\theta u(x)}dx < \infty \right\} \tag{1.8}$$

for $\gamma \in (c, d)$. Then, it is shown that for any $\gamma_1, \gamma_2 \in (c, d)$ with $\gamma_1 < \gamma_2$, $\Theta(\gamma_1) \subset \Theta(\gamma_2)$. Assume that for any $\gamma \in (c, d)$, $\Theta \equiv \Theta(\gamma)$ is a non-empty open interval. A family $\mathscr{P}_o := \{P_{\theta,\gamma} \mid \theta \in \Theta, \gamma \in (c, d)\}$ of distributions $P_{\theta,\gamma}$ with a natural parameter θ and a truncation parameter γ is called a one-sided truncated exponential family (oTEF) of distributions; to be precise, $\mathscr{P}_o$ may be called a lower-truncated exponential family of distributions. Suppose that a random variable X is distributed according to $P_{\theta,\gamma}$ in $\mathscr{P}_o$. For example, a lower-truncated exponential, a lower-truncated normal and Pareto distributions belong to the oTEF. If γ is known, then $\mathscr{P}_o$ is a regular exponential family of distributions. Letting $Y = -X$, we easily see that Y constitutes an upper-truncated exponential family of distributions.

For a oTEF $\mathscr{P}_o$ with the density (1.7), the vector (θ_1, θ_2) and ϑ in Sect. 1.1 are regarded as (θ, γ) and γ. The maximum likelihood estimation of θ is discussed when γ is a nuisance parameter in Chap. 2, and the estimation of a truncation parameter γ is also done when θ is a nuisance parameter in Chaps. 4 and 6.

1.3 Two-Sided Truncated Exponential Family

In a similar way to the previous section, we consider a distribution $P_{\theta,\gamma,\nu}$, having density

$$f(x;\theta,\gamma,\nu) = \begin{cases} \frac{a(x)e^{\theta u(x)}}{b(\theta,\gamma,\nu)} & \text{for } c < \gamma \le x \le \nu < d, \\ 0 & \text{otherwise} \end{cases} \tag{1.9}$$

with respect to the Lebesgue measure, where $-\infty \le c < d \le \infty$, $a(\cdot)$ is a nonnegative-valued and continuous almost surely, and $u(\cdot)$ is absolutely continuous with $du(x)/dx \not\equiv 0$ over the interval (γ, ν) for $\gamma, \nu \in (c, d)$ and $\gamma < \nu$. Let

$$\Theta(\gamma,\nu) := \left\{\theta \,\middle|\, 0 < b(\theta,\gamma,\nu) := \int_\gamma^\nu a(x)e^{\theta u(x)}dx < \infty\right\} \tag{1.10}$$

for $\gamma, \nu \in (c, d)$ and $\gamma < \nu$. Assume that for $\gamma, \nu \in (c, d)$ with $\gamma < \nu$, $\Theta \equiv \Theta(\gamma, \nu)$ is a non-empty open interval. A family $\mathscr{P}_t := \{P_{\theta,\gamma,\nu} \mid \theta \in \Theta,\ \gamma, \nu \in (c, d), \gamma < \nu\}$ of distributions $P_{\theta,\gamma,\nu}$ with a natural parameter θ and truncation parameters γ, ν is called a two-sided truncated exponential family (tTEF) of distributions. Here, γ and ν are said to be lower and upper truncation parameters. If γ and ν are known, then $\mathscr{P}_t$ is a regular exponential family of distributions. For example, a two-sided truncated exponential, a two-sided truncated normal and an upper-truncated Pareto distributions belong to the tTEF. The maximum likelihood estimation of θ is discussed when γ and ν are nuisance parameters in Chap. 3, and the estimation of a truncation parameters is also done when θ and another truncation parameter are nuisance parameters in Chap. 5.

References

Akahira, M. (1975). Asymptotic theory for estimation of location in non-regular cases, I: Order of convergence of consistent estimators. *Reports of Statistical Application Research JUSE*, *22*, 8–26.

Akahira, M. (1981). On asymptotic deficiency of estimators. *Australian Journal of Statistics*, *23*, 67–72.

Akahira, M. (1986). *The structure of asymptotic deficiency of estimators*. Queen's Papers in Pure and Applied Mathematics (Vol. 75). Kingston, Canada: Queen's University Press.

Akahira, M. (1992). Higher order asymptotics and asymptotic deficiency of estimators. *Selecta Statistica Canadiana*, *8*, 1–36.

Akahira, M. (1995). The amount of information and the bound for the order of consistency for a location parameter family of densities. In: R. Fritsch, M. Behara, & R. G. Lintz, (Eds.) *Proceedings of the 2nd Gauss symposium. Conference B: statistical sciences* (pp. 303–311). Berlin: de Gruyter.

Akahira, M., & Takeuchi, K. (1981). *Asymptotic efficiency of statistical estimators: Concepts and higher order asymptotic efficiency* (Vol. 7). Lecture notes in statistics. New York: Springer.

Akahira, M., & Takeuchi, K. (1982). On asymptotic deficiency of estimators in pooled samples in the presence of nuisance parameters. *Statistics & Decisions*, *1*, 17–38.

Akahira, M., & Takeuchi, K. (1987). On the definition of asymptotic expectation. In: I. B. MacNeill and G. J. Umphrey (Eds.), *Foundations of statistical inference*, (pp. 199–208). D. Reidel Publishing Company (Also included In: *"Joint Statistical Papers of Akahira and Takeuchi,"* World Scientific, New Jersey, 2003).

Akahira, M., & Takeuchi, K. (1995). *Non-regular statistical estimation* (Vol. 107). Lecture notes in statistics. New York: Springer.

Bar-Lev, S. K. (1984). Large sample properties of the MLE and MCLE for the natural parameter of a truncated exponential family. *Annals of the Institute of Statistical Mathematics*, *36*, Part A, 217–222.

Barndorff-Nielsen, O. E., & Cox, D. R. (1994). *Inference and asymptotics*. London: Chapman & Hall.

Cox, D. R., & Reid, N. (1987). Parameter orthogonality and approximate conditional inference (with discussion). *Journal of the Royal Statistical Society Series B*, *49*, 1–39.

Hodges, J. L., & Lehmann, E. L. (1970). Deficiency. *The Annals of Mathematical Statistics*, *41*, 783–801.

Lehmann, E. L. (1999). *Elements of large-sample theory*. New York: Springer.

Chapter 2
Maximum Likelihood Estimation of a Natural Parameter for a One-Sided TEF

For a one-sided truncated exponential family (oTEF) of distributions with a natural parameter θ and a truncation parameter γ as a nuisance parameter, the maximum likelihood estimators (MLEs) $\hat{\theta}_{ML}^{\gamma}$ and $\hat{\theta}_{ML}$ of θ for known γ and unknown γ and the maximum conditional likelihood estimator (MCLE) $\hat{\theta}_{MCL}$ of θ are asymptotically compared up to the second order.

2.1 Introduction

In the presence of nuisance parameters, the asymptotic loss of the maximum likelihood estimator of an interest parameter was discussed by Akahira and Takeuchi (1982) and Akahira (1986) under suitable regularity conditions from the viewpoint of higher order asymptotics. On the other hand, in statistical estimation in multiparameter cases, the conditional likelihood method is well known as a way of eliminating nuisance parameters (see, e.g., Basu 1997). The consistency, asymptotic normality, and asymptotic efficiency of the MCLE were discussed by Andersen (1970), Huque and Katti (1976), Bar-Lev and Reiser (1983), Bar-Lev (1984), Liang (1984), and others. Further, in higher order asymptotics, asymptotic properties of the MCLE of an interest parameter in the presence of nuisance parameters were also discussed by Cox and Reid (1987) and Ferguson (1992) in the regular case. However, in the non-regular case when the regularity conditions do not necessarily hold, the asymptotic comparison of asymptotically efficient estimators has not been discussed enough in the presence of nuisance parameters in higher order asymptotics yet.

For a truncated exponential family of distributions which is regarded as a typical non-regular case, we consider a problem of estimating a natural parameter θ in the presence of a truncation parameter γ as a nuisance parameter. Let $\hat{\theta}_{ML}^{\gamma}$ and $\hat{\theta}_{ML}$ be

M. Akahira, *Statistical Estimation for Truncated Exponential Families*,
JSS Research Series in Statistics, DOI 10.1007/978-981-10-5296-5_2

the MLEs of θ based on a sample of size n when γ is known and γ is unknown, respectively. Let $\hat{\theta}_{MCL}$ be the MCLE of θ. Then, it was shown by Bar-Lev (1984) that the MLEs $\hat{\theta}_{ML}^{\gamma}$,$\hat{\theta}_{ML}$ and the MCLE $\hat{\theta}_{MCL}$ have the same asymptotic normal distribution, hence they are shown to be asymptotically equivalent in the sense of having the same asymptotic variance. A similar result can be derived from the stochastic expansions of the MLEs $\hat{\theta}_{ML}^{\gamma}$ and $\hat{\theta}_{ML}$ in Akahira and Ohyauchi (2012). But, $\hat{\theta}_{ML}^{\gamma}$ for known γ may be asymptotically better than $\hat{\theta}_{ML}$ for unknown γ in the higher order, because $\hat{\theta}_{ML}^{\gamma}$ has the full information on γ. Otherwise, the existence of a truncation parameter γ as a nuisance parameter is meaningless. So, it is a quite interesting problem to compare asymptotically them up to the higher order.

In this chapter, following mostly the paper by Akahira (2016), we compare them up to the second order, i.e., the order of n^{-1}, in the asymptotic variance. We show that a bias-adjusted MLE $\hat{\theta}_{ML^*}$ and $\hat{\theta}_{MCL}$ are second order asymptotically equivalent, but they are asymptotically worse than $\hat{\theta}_{ML}^{\gamma}$ in the second order. We thus calculate the second-order asymptotic losses on the asymptotic variance among them. Several examples are also given.

2.2 Preliminaries

Suppose that $X_1, X_2, \dots, X_n, \dots$ is a sequence of independent and identically distributed (i.i.d.) random variables according to $P_{\theta,\gamma}$ in a oTEF $\mathscr{P}_o$ with the density (1.7). In Bar-Lev (1984), the asymptotic behavior of the MLE $\hat{\theta}_{ML}$ and MCLE $\hat{\theta}_{MCL}$ of a parameter θ in the presence of γ as a nuisance parameter was compared and also done with that of the MLE $\hat{\theta}_{ML}^{\gamma}$ of θ when γ was known. As the result, it was shown there that, for a sample of size $n(\geq 2)$, the $\hat{\theta}_{ML}$ and $\hat{\theta}_{MCL}$ of θ existed with probability 1 and were given as the unique roots of the appropriate maximum likelihood equations. These two estimators were also shown to be strongly consistent for θ with the limiting distribution which coincides with that of the MLE $\hat{\theta}_{ML}^{\gamma}$ of θ when γ was known. Denote a random vector $(X_1, \dots, X_n)$ by $\boldsymbol{X}$ and let $X_{(1)} \leq \cdots \leq X_{(n)}$ be the corresponding order statistics of a random vector $\boldsymbol{X}$. Then, the density (1.7) is considered to belong to a regular exponential family of distributions with a natural parameter θ for any fixed γ, hence $\log b(\theta, \gamma)$ is strictly convex and infinitely differentiable in $\theta \in \Theta$ and

$$\lambda_j(\theta, \gamma) := \frac{\partial^j}{\partial \theta^j} \log b(\theta, \gamma) \tag{2.1}$$

is the j th cumulant corresponding to (1.7) for $j = 1, 2, \dots$.

In the subsequent sections, we obtain the stochastic expansions of $\hat{\theta}^{\gamma}_{ML}$, $\hat{\theta}_{ML}$, and $\hat{\theta}_{MCL}$ up to the second order, i.e., $o_p(n^{-1})$. We get their second-order asymptotic variances and derive the second-order asymptotic losses on the asymptotic variance among them. The proofs of theorems are located in Appendixes A1 and A2.

2.3 MLE $\hat{\theta}^{\gamma}_{ML}$ of a Natural Parameter θ When a Truncation Parameter γ is Known

For given $\boldsymbol{x} = (x_1, \dots, x_n)$ satisfying $\gamma \le x_{(1)} := \min_{1\le i\le n} x_i$ and $x_{(n)} := \max_{1\le i\le n} x_i < d$, the likelihood function of θ is given by

$$L^{\gamma}(\theta; \boldsymbol{x}) := \frac{1}{b^n(\theta,\gamma)} \left\{\prod_{i=1}^{n} a(x_i)\right\} \exp\left\{\theta \sum_{i=1}^{n} u(x_i)\right\}.$$

Then, the likelihood equation is

$$\frac{1}{n}\sum_{i=1}^{n} u(X_i) - \lambda_1(\theta,\gamma) = 0. \tag{2.2}$$

Since there exists a unique solution of Eq. (2.2) with respect to θ, we denote it by $\hat{\theta}^{\gamma}_{ML}$ which is the MLE of θ (see, e.g., Barndorff-Nielsen (1978) and Bar-Lev (1984)). Let $\lambda_i = \lambda_i(\theta,\gamma)$ $(i = 2, 3, 4)$ and put

$$Z_1 := \frac{1}{\sqrt{\lambda_2 n}} \sum_{i=1}^{n} \{u(X_i) - \lambda_1\}, \quad U_{\gamma} := \sqrt{\lambda_2 n}\left(\hat{\theta}^{\gamma}_{ML} - \theta\right).$$

Then, we have the following.

Theorem 2.3.1 *For the oTEF $\mathscr{P}_o$ of distributions with densities of the form (1.7) with a natural parameter θ and a truncation parameter γ, let $\hat{\theta}^{\gamma}_{ML}$ be the MLE of θ when γ is known. Then, the stochastic expansion of U_{γ} is given by*

$$U_{\gamma} = Z_1 - \frac{\lambda_3}{2\lambda_2^{3/2}\sqrt{n}} Z_1^2 + \frac{1}{2n}\left(\frac{\lambda_3^2}{\lambda_2^3} - \frac{\lambda_4}{3\lambda_2^2}\right) Z_1^3 + O_p\left(\frac{1}{n\sqrt{n}}\right),$$

and the second-order asymptotic mean and variance are given by

$$E_\theta\left(U_\gamma\right) = -\frac{\lambda_3}{2\lambda_2^{3/2}\sqrt{n}} + O\left(\frac{1}{n\sqrt{n}}\right),$$

$$V_\theta\left(U_\gamma\right) = 1 + \frac{1}{n}\left(\frac{5\lambda_3^2}{2\lambda_2^3} - \frac{\lambda_4}{\lambda_2^2}\right) + O\left(\frac{1}{n\sqrt{n}}\right),$$

respectively.

Since $U_\gamma = Z_1 + o_p(1)$, it is seen that U_γ is asymptotically normal with mean 0 and variance 1, which coincides with the result of Bar-Lev (1984).

2.4 Bias-Adjusted MLE $\hat{\theta}_{ML^*}$ of θ When γ is Unknown

For given $\boldsymbol{x} = (x_1, \ldots, x_n)$ satisfying $\gamma \le x_{(1)}$ and $x_{(n)} < d$, the likelihood function of θ and γ is given by

$$L(\theta, \gamma; \boldsymbol{x}) = \frac{1}{b^n(\theta, \gamma)} \left\{\prod_{i=1}^{n} a(x_i)\right\} \exp\left\{\theta \sum_{i=1}^{n} u(x_i)\right\}. \tag{2.3}$$

Let $\hat{\theta}_{ML}$ and $\hat{\gamma}_{ML}$ be the MLEs of θ and γ, respectively. From (2.3), it is seen that $\hat{\gamma}_{ML} = X_{(1)}$ and $L(\hat{\theta}_{ML}, X_{(1)}; \boldsymbol{X}) = \sup_{\theta\in\Theta} L(\theta, X_{(1)}; \boldsymbol{X})$, hence $\hat{\theta}_{ML}$ satisfies the likelihood equation

$$0 = \frac{1}{n}\sum_{i=1}^{n} u(X_i) - \lambda_1(\hat{\theta}_{ML}, X_{(1)}), \tag{2.4}$$

where $\boldsymbol{X} = (X_1, \cdots, X_n)$. Let $\lambda_2 = \lambda_2(\theta, \gamma)$ and put $\hat{U} := \sqrt{\lambda_2 n}(\hat{\theta}_{ML} - \theta)$ and $T_{(1)} := n(X_{(1)} - \gamma)$. Then, we have the following.

Theorem 2.4.1 *For the oTEF $\mathscr{P}_o$ of distributions with densities of the form (1.7) with a natural parameter θ and a truncation parameter γ, let $\hat{\theta}_{ML}$ be the MLE of θ when γ is unknown, and $\hat{\theta}_{ML^*}$ be a bias-adjusted MLE such that $\hat{\theta}_{ML}$ has the same asymptotic bias as that of $\hat{\theta}_{ML}^{\gamma}$, i.e.,*

$$\hat{\theta}_{ML^*} = \hat{\theta}_{ML} + \frac{1}{k(\hat{\theta}_{ML}, X_{(1)})\lambda_2(\hat{\theta}_{ML}, X_{(1)})n} \left\{\frac{\partial\lambda_1}{\partial\gamma}\left(\hat{\theta}_{ML}, X_{(1)}\right)\right\}, \tag{2.5}$$

where $k(\theta, \gamma) := a(\gamma)e^{\theta u(\gamma)}/b(\theta, \gamma)$. Then, the stochastic expansion of $\hat{U}^* := \sqrt{\lambda_2 n}(\hat{\theta}_{ML^*} - \theta)$ is given by

$$\hat{U}^* = \hat{U} + \frac{1}{k\sqrt{\lambda_2 n}}\left(\frac{\partial\lambda_1}{\partial\gamma}\right) - \frac{1}{k\lambda_2 n}\left\{\delta + \frac{1}{k}\left(\frac{\partial k}{\partial\theta}\frac{\partial\lambda_1}{\partial\gamma}\right)\right\} Z_1 + O_p\left(\frac{1}{n\sqrt{n}}\right),$$

where $k = k(\theta, \gamma)$,

$$\delta = \frac{\lambda_3}{\lambda_2}\left(\frac{\partial \lambda_1}{\partial \gamma}\right) - \frac{\partial \lambda_2}{\partial \gamma},$$

$$\hat{U} = Z_1 - \frac{\lambda_3}{2\lambda_2^{3/2}\sqrt{n}} Z_1^2 - \frac{1}{\sqrt{\lambda_2 n}}\left(\frac{\partial \lambda_1}{\partial \gamma}\right) T_{(1)} + \frac{\delta}{\lambda_2 n} Z_1 T_{(1)} + \frac{1}{2n}\left(\frac{\lambda_3^2}{\lambda_2^3} - \frac{\lambda_4}{3\lambda_2^2}\right) Z_1^3 + O_p\left(\frac{1}{n\sqrt{n}}\right),$$

and the second-order asymptotic mean and variance are given by

$$E_{\theta,\gamma}(\hat{U}^*) = -\frac{\lambda_3}{2\lambda_2^{3/2}\sqrt{n}} + O\left(\frac{1}{n\sqrt{n}}\right),$$

$$V_{\theta,\gamma}(\hat{U}^*) = 1 + \frac{1}{n}\left(\frac{5\lambda_3^2}{2\lambda_2^3} - \frac{\lambda_4}{\lambda_2^2}\right) + \frac{1}{\lambda_2 n}\{u(\gamma) - \lambda_1\}^2 + O\left(\frac{1}{n\sqrt{n}}\right),$$

respectively.

Since $\hat{U} = \hat{U}^* = Z_1 + o_p(1)$, it is seen that $\hat{U}$ and $\hat{U}^*$ are asymptotically normal with mean 0 and variance 1 in the first order, which coincides with the result of Bar-Lev (1984). But, it is noted from Theorems 2.3.1, and 2.4.1 that there is a difference between $V_\theta(U_\gamma)$ and $V_{\theta,\gamma}(\hat{U}^*)$ in the second order, i.e., the order n^{-1}, which is discussed in Sect. 2.6. It is also remarked that the asymptotic distribution of $T_{(1)}$ is exponential in the first order and given up to the second order (see Lemma 2.9.1 in later Appendix A1).

2.5 MCLE $\hat{\theta}_{MCL}$ of θ When γ is Unknown

First, it is seen from (1.7) that there exists a random permutation, say $Y_2, \cdots, Y_n$ of the $(n-1)!$ permutations of $(X_{(2)}, \ldots, X_{(n)})$ such that conditionally on $X_{(1)} = x_{(1)}$, $Y_2, \ldots, Y_n$ are i.i.d. random variables according to a distribution with density

$$g(y; \theta, x_{(1)}) = \frac{a(y)e^{\theta u(y)}}{b(\theta, x_{(1)})} \quad \text{for } x_{(1)} \le y < d$$

with respect to the Lebesgue measure (see Quesenberry (1975) and Bar-Lev (1984)). For given $X_{(1)} = x_{(1)}$, the conditional likelihood function of θ for $\mathbf{y} = (y_2, \ldots, y_n)$ satisfying $x_{(1)} \le y_i < d$ $(i = 2, \ldots, n)$ is

$$L(\theta; \mathbf{y}|x_{(1)}) = \frac{1}{b^{n-1}(\theta, x_{(1)})}\left\{\prod_{i=2}^{n} a(y_i)\right\} \exp\left\{\theta \sum_{i=2}^{n} u(y_i)\right\}.$$

Then, the likelihood equation is

$$\frac{1}{n-1}\sum_{i=2}^{n} u(y_i) - \lambda_1(\theta, x_{(1)}) = 0. \tag{2.6}$$

Since there exists a unique solution on θ of (2.6), we denote it by the MCLE $\hat{\theta}_{MCL}$, i.e., the value of θ for which $L(\theta; \mathbf{y}|x_{(1)})$ attains supremum. Let $\tilde{\lambda}_i := \lambda_i(\theta, x_{(1)})$ $(i = 1, 2, 3, 4)$ and put

$$\tilde{Z}_1 := \frac{1}{\sqrt{\tilde{\lambda}_2(n-1)}}\sum_{i=2}^{n}\left\{u(Y_i) - \tilde{\lambda}_1\right\}, \quad \tilde{U}_0 := \sqrt{\lambda_2 n}\left(\hat{\theta}_{MCL} - \theta\right).$$

Then, we have the following.

Theorem 2.5.1 *For a oTEF $\mathscr{P}_o$ of distributions with densities of the form (1.7) with a natural parameter θ and a truncation parameter γ, let $\hat{\theta}_{MCL}$ be the MCLE of θ when γ is unknown. Then, the stochastic expansion of $\tilde{U}_0$ is given by*

$$\tilde{U}_0 = \tilde{Z}_1 - \frac{\tilde{\lambda}_3}{2\tilde{\lambda}_2^{3/2}\sqrt{n}}\tilde{Z}_1^2 + \frac{1}{2n}\left\{1 - \frac{1}{\lambda_2}\left(\frac{\partial\lambda_2}{\partial\gamma}\right)T_{(1)}\right\}\tilde{Z}_1$$
$$+\frac{1}{2n}\left(\frac{\tilde{\lambda}_3^2}{\tilde{\lambda}_2^3} - \frac{\tilde{\lambda}_4}{3\tilde{\lambda}_2^2}\right)\tilde{Z}_1^3 + O_p\left(\frac{1}{n\sqrt{n}}\right),$$

and the second-order asymptotic mean and variance are given by

$$E_{\theta,\gamma}\left(\tilde{U}_0\right) = -\frac{\lambda_3}{2\lambda_2^{3/2}\sqrt{n}} + O\left(\frac{1}{n\sqrt{n}}\right),$$
$$V_{\theta,\gamma}\left(\tilde{U}_0\right) = 1 + \frac{1}{n}\left(\frac{5\lambda_3^2}{2\lambda_2^3} - \frac{\lambda_4}{\lambda_2^2}\right) + \frac{1}{\lambda_2 n}\{u(\gamma) - \lambda_1\}^2 + O\left(\frac{1}{n\sqrt{n}}\right).$$

Remark 2.5.1 From Theorems 2.4.1, and 2.5.1, it is seen that the second-order asymptotic mean and variance of $\tilde{U}_0$ are the same as those of $\hat{U}^* = \sqrt{\lambda_2 n}(\hat{\theta}_{ML^*} - \theta)$. It is noted that $\hat{\theta}_{MCL}$ has an advantage over $\hat{\theta}_{ML}$ in the sense of no need of the bias-adjustment.

Remark 2.5.2 As is seen from Theorems 2.3.1, 2.4.1, and 2.5.1, the first terms of order $1/n$ in $V_\theta(U_\gamma)$, $V_{\theta,\gamma}(\hat{U}^*)$, and $V_{\theta,\gamma}(\tilde{U}_0)$ result from the regular part of the density (1.7), which coincides with the fact that the distribution with (1.7) is considered to belong to a regular exponential family of distributions when γ is known. The second terms of order $1/n$ in $V_{\theta,\gamma}(\hat{U}^*)$ and $V_{\theta,\gamma}(\tilde{U}_0)$ follow from the non-regular (i.e., truncation) part of (1.7) when γ is unknown, which means a ratio of the variance $\lambda_2 = V_{\theta,\gamma}(u(X)) = E_{\theta,\gamma}[\{u(X) - \lambda_1\}^2]$ to the distance $\{u(\gamma) - \lambda_1\}^2$ from the mean λ_1 of $u(X)$ to $u(x)$ at $x = \gamma$.

2.6 Second-Order Asymptotic Comparison Among $\hat{\theta}_{ML}^{\gamma}$, $\hat{\theta}_{ML^*}$, and $\hat{\theta}_{MCL}$

From the results in the previous sections, we can asymptotically compare the estimators $\hat{\theta}_{ML}^{\gamma}$, $\hat{\theta}_{ML^*}$, and $\hat{\theta}_{MCL}$ using their second-order asymptotic variances as follows.

Theorem 2.6.1 *For a oTEF $\mathscr{P}_o$ of distributions with densities of the form (1.7) with a natural parameter θ and a truncation parameter γ, let $\hat{\theta}_{ML}^{\gamma}$, $\hat{\theta}_{ML^*}$, and $\hat{\theta}_{MCL}$ be the MLE of θ when γ is known, the bias-adjusted MLE of θ when γ is unknown and the MCLE of θ when γ is unknown, respectively. Then, the bias-adjusted MLE $\hat{\theta}_{ML}^{*}$ and the MCLE $\hat{\theta}_{MCL}$ are second order asymptotically equivalent in the sense that*

$$d_n(\hat{\theta}_{ML^*}, \hat{\theta}_{MCL}) := n\left\{V_{\theta,\gamma}(\hat{U}^*) - V_{\theta,\gamma}(\tilde{U}_0)\right\} = o(1) \tag{2.7}$$

as $n \to \infty$ and they are second order asymptotically worse than $\hat{\theta}_{ML}^{\gamma}$ with the following second-order asymptotic losses of $\hat{\theta}_{ML^*}$and $\hat{\theta}_{MCL}$ relative to $\hat{\theta}_{ML}^{\gamma}$:

$$d_n(\hat{\theta}_{ML^*}, \hat{\theta}_{ML}^{\gamma}) := n\left\{V_{\theta,\gamma}(\hat{U}^*) - V_{\theta}(U_{\gamma})\right\} = \frac{\{u(\gamma) - \lambda_1\}^2}{\lambda_2} + o(1), \tag{2.8}$$

$$d_n(\hat{\theta}_{MCL}, \hat{\theta}_{ML}^{\gamma}) := n\left\{V_{\theta,\gamma}(\tilde{U}_0) - V_{\theta}(U_{\gamma})\right\} = \frac{\{u(\gamma) - \lambda_1\}^2}{\lambda_2} + o(1) \tag{2.9}$$

as $n \to \infty$, respectively.

The proof is straightforward from Theorems 2.3.1, 2.4.1, and 2.5.1.

Remark 2.6.1 It is seen from (1.6) and (2.8) that the ratio of the asymptotic variance of $\hat{U}^*$ to that of U_{γ} is given by

$$R_n(\hat{\theta}_{ML^*}, \hat{\theta}_{ML}^{\gamma}) = 1 + \frac{\{u(\gamma) - \lambda_1\}^2}{\lambda_2 n} + o\left(\frac{1}{n}\right),$$

and similarly from (1.6) and (2.9)

$$R_n(\hat{\theta}_{MCL}, \hat{\theta}_{ML}^{\gamma}) = 1 + \frac{\{u(\gamma) - \lambda_1\}^2}{\lambda_2 n} + o\left(\frac{1}{n}\right).$$

From the consideration of models in Sect. 1.1, using (1.5), (1.6), and (2.8) we see that the difference between the asymptotic models $M(\hat{\theta}_{ML^*}, \gamma)$ and $M(\hat{\theta}_{ML}^{\gamma}, \gamma)$ is given by $d_n(\hat{\theta}_{ML^*}, \hat{\theta}^{\gamma}_{ML})$ or $R_n(\hat{\theta}_{ML^*}, \hat{\theta}^{\gamma}_{ML})$ up to the second order, through the MLE of θ. In a similar way to the above, the difference between $M(\hat{\theta}_{MCL}, \gamma)$ and $M(\hat{\theta}^{\gamma}_{MCL}, \gamma)$ is given by $d_n(\hat{\theta}_{MCL}, \hat{\theta}^{\gamma}_{ML})$ or $R_n(\hat{\theta}_{MCL}, \hat{\theta}^{\gamma}_{ML})$ up to the second order.

2.7 Examples

Examples on the second-order asymptotic losses of the estimators are given for a lower-truncated exponential, a lower-truncated normal, and Pareto, a lower-truncated beta and a lower-truncated Erlang type distributions.

Example 2.7.1 **(Lower-truncated exponential distribution)** Let $c = -\infty, d = \infty$, $a(x) = 1$, and $u(x) = -x$ for $-\infty < \gamma \le x < \infty$ in the density (1.7). Since $b(\theta, \gamma) = e^{-\theta\gamma}/\theta$ for $\theta \in \Theta = (0, \infty)$, it follows from (2.1) that

$$\lambda_1 = \frac{\partial}{\partial\theta}\log b(\theta, \gamma) = -\gamma - \frac{1}{\theta},$$

$$\lambda_2 = \frac{\partial^2}{\partial\theta^2}\log b(\theta, \gamma) = \frac{1}{\theta^2}, \quad k(\theta, \gamma) = \theta.$$

From (2.2) and (2.4)–(2.6), we have

$$\hat{\theta}^{\gamma}_{ML} = 1/(\bar{X} - \gamma), \quad \hat{\theta}_{ML} = 1/(\bar{X} - X_{(1)}),$$

$$\hat{\theta}_{ML^*} = \hat{\theta}_{ML} - \frac{1}{n}\hat{\theta}_{ML}, \quad \hat{\theta}_{MCL} = 1\Bigg/\left(\frac{1}{n-1}\sum_{i=2}^{n} X_{(i)} - X_{(1)}\right).$$

Note that $\hat{\theta}_{ML^*} = \hat{\theta}_{MCL}$. In this case, the first part in Theorem 2.6.1 is trivial, since $d_n(\hat{\theta}_{ML^*}, \hat{\theta}_{MCL}) = 0$. From Theorem 2.6.1, we obtain the second-order asymptotic loss

$$d_n(\hat{\theta}_{ML^*}, \hat{\theta}^{\gamma}_{ML}) = d_n(\hat{\theta}_{MCL}, \hat{\theta}^{\gamma}_{ML}) = 1 + o(1)$$

as $n \to \infty$. Note that the loss is independent of γ up to the order $o(1)$. From Remark 2.6.1, we have the ratio

$$R_n(\hat{\theta}_{ML^*}, \hat{\theta}^{\gamma}_{ML}) = R_n(\hat{\theta}_{MCL}, \hat{\theta}^{\gamma}_{ML}) = 1 + \frac{1}{n} + o\left(\frac{1}{n}\right).$$

In this case, we have the uniformly minimum variance unbiased (UMVU) estimator

$$\hat{\theta}^{\gamma}_{UMVU} = (n-1)\Bigg/\left(\sum_{i=1}^{n} X_i - n\gamma\right),$$

where γ is known (see Voinov and Nikulin 1993). Then,

$$\hat{\theta}_{ML}^{\gamma} = \left(1 + \frac{1}{n-1}\right)\hat{\theta}_{UMVU}^{\gamma},$$

hence $\hat{\theta}_{ML}^{\gamma}$ is not unbiased for any fixed n. When γ is unknown, we obtain the UMVU estimator

$$\hat{\theta}_{UMVU} = (n-2) \bigg/ \left\{\sum_{i=2}^{n} X_{(i)} - (n-1)X_{(1)}\right\}$$

which is derived from the formula in the lower-truncated exponential distribution of a general type discussed by Lwin (1975) and Voinov and Nikulin (1993). Then,

$$\hat{\theta}_{ML^*} = \hat{\theta}_{MCL} = \left(1 + \frac{1}{n-2}\right)\hat{\theta}_{UMVU},$$

hence $\hat{\theta}_{ML^*}$ and $\hat{\theta}_{MCL}$ are not unbiased for any fixed n. Note that $\hat{\theta}_{ML^*}$ and $\hat{\theta}_{MCL}$ are asymptotically compared with $\hat{\theta}_{ML}^{\gamma}$ after such a bias adjustment that $\hat{\theta}_{ML}$ has the same asymptotic bias given in Theorem 2.3.1 as $\hat{\theta}_{ML}^{\gamma}$. Since $\lambda_2 = 1/\theta^2$,

$$V_{\theta}(\hat{\theta}_{UMVU}^{\gamma}) = \frac{\theta^2}{n-2}, \quad V_{\theta,\gamma}(\hat{\theta}_{UMVU}) = \frac{\theta^2}{n-3},$$

we have the second-order asymptotic loss

$$\begin{aligned} d_n(\hat{\theta}_{UMVU}, \hat{\theta}_{UMVU}^{\gamma}) &= n\left\{V_{\theta,\gamma}\left(\frac{\sqrt{n}}{\theta}(\hat{\theta}_{UMVU} - \theta)\right) - V_{\theta}\left(\frac{\sqrt{n}}{\theta}(\hat{\theta}_{UMVU}^{\gamma} - \theta)\right)\right\} \\ &= \frac{n^2}{(n-2)(n-3)} = 1 + o(1) \end{aligned}$$

as $n \to \infty$.

Example 2.7.2 (**Lower-truncated normal distribution**) Let $c = -\infty$, $d = \infty$, $a(x) = e^{-x^2/2}$ and $u(x) = x$ for $-\infty < \gamma \le x < \infty$ in the density (1.7). Since

$$b(\theta, \gamma) = \sqrt{2\pi} e^{\theta^2/2} \Phi(\theta - \gamma)$$

for $\theta \in \Theta = (-\infty, \infty)$, it follows from (2.1) and Theorem 2.4.1 that

$$\lambda_1(\theta, \gamma) = \theta + \rho(\theta - \gamma), \quad \frac{\partial \lambda_1}{\partial \gamma}(\theta, \gamma) = (\theta - \gamma)\rho(\theta - \gamma) + \rho^2(\theta - \gamma),$$

$$\lambda_2(\theta, \gamma) = 1 - (\theta - \gamma)\rho(\theta - \gamma) - \rho^2(\theta - \gamma), \quad k(\theta, \gamma) = \rho(\theta - \gamma),$$

Table 2.1 Values of $d_n(\hat{\theta}_{ML^*}, \hat{\theta}_{ML}^{\gamma})$ and $R_n(\hat{\theta}_{ML^*}, \hat{\theta}_{ML}^{\gamma})$ for $\tau = \theta - \gamma = 2, 4, 6$

τ	$d_n(\hat{\theta}_{ML^*}, \hat{\theta}_{ML}^{\gamma})$	$R_n(\hat{\theta}_{ML^*}, \hat{\theta}_{ML}^{\gamma})$
2	$4.7640 + o(1)$	$1 + \dfrac{4.7640}{n} + o\left(\dfrac{1}{n}\right)$
4	$9.1486 + o(1)$	$1 + \dfrac{9.1486}{n} + o\left(\dfrac{1}{n}\right)$
6	$16.0096 + o(1)$	$1 + \dfrac{16.0096}{n} + o\left(\dfrac{1}{n}\right)$

where $\rho(t) := \phi(t)/\Phi(t)$ with $\Phi(x) = \int_{-\infty}^{x} \phi(t)dt$ and $\phi(t) = (1/\sqrt{2\pi})e^{-t^2/2}$ for $-\infty < t < \infty$. Then, it follows from (2.2), (2.4), and (2.6) that the solutions of θ of the following equations

$$\theta + \rho(\theta - \gamma) = \bar{X}, \quad \theta + \rho(\theta - X_{(1)}) = \bar{X},$$

$$\theta + \rho(\theta - X_{(1)}) = \frac{1}{n-1}\sum_{i=2}^{n} X_{(i)}$$

become $\hat{\theta}_{ML}^{\gamma}$, $\hat{\theta}_{ML}$, and $\hat{\theta}_{MCL}$, respectively, where $\bar{X} = (1/n)\sum_{i=1}^{n} X_i$. From (2.5), the bias-adjusted MLE is given by

$$\hat{\theta}_{ML^*} = \hat{\theta}_{ML} + \frac{\hat{\theta}_{ML} - X_{(1)} + \rho(\hat{\theta}_{ML} - X_{(1)})}{1 - (\hat{\theta}_{ML} - X_{(1)})\rho(\hat{\theta}_{ML} - X_{(1)}) - \rho^2(\hat{\theta}_{ML} - X_{(1)})}.$$

From Theorem 2.6.1, we obtain the second-order asymptotic losses

$$d_n(\hat{\theta}_{ML^*}, \hat{\theta}_{MCL}) = o(1),$$
$$d_n(\hat{\theta}_{ML^*}, \hat{\theta}_{ML}^{\gamma}) = d_n(\hat{\theta}_{MCL}, \hat{\theta}_{ML}^{\gamma}) = \frac{\{\theta - \gamma + \rho(\theta - \gamma)\}^2}{1 - (\theta - \gamma)\rho(\theta - \gamma) - \rho^2(\theta - \gamma)} + o(1)$$

as $n \to \infty$.

When $\tau := \theta - \gamma = 2, 4, 6$, the value of second-order asymptotic loss $d_n(\hat{\theta}_{ML^*}, \hat{\theta}_{ML}^{\gamma})$ and the ratio $R_n(\hat{\theta}_{ML^*}, \hat{\theta}_{ML}^{\gamma})$ up to the order $1/n$ are obtained from (2.8) and Remark 2.6.1 (see Table 2.1 and Fig. 2.1).

Example 2.7.3 (**Pareto distribution**) Let $c = 0$, $d = \infty$, $a(x) = 1/x$, and $u(x) = -\log x$ for $0 < \gamma \le x < \infty$ in the density (1.7). Then, $b(\theta, \gamma) = 1/(\theta\gamma^{\theta})$ for $\theta \in \Theta = (0, \infty)$. Letting $t = \log x$ and $\gamma_0 = \log \gamma$, we see that (1.7) becomes

$$f(t; \theta, \gamma_0) = \begin{cases} \theta e^{\theta\gamma_0} e^{-\theta t} & \text{for } t \ge \gamma_0, \\ 0 & \text{for } t < \gamma_0. \end{cases}$$

Hence, the Pareto case is reduced to the truncated exponential one in Example 2.7.1. Replacing $\bar{X}$ and $X_{(i)}$ $(i = 1, \ldots, n)$ by $\overline{\log X} := (1/n)\sum_{i=1}^{n} \log X_i$ and $\log X_{(i)}$

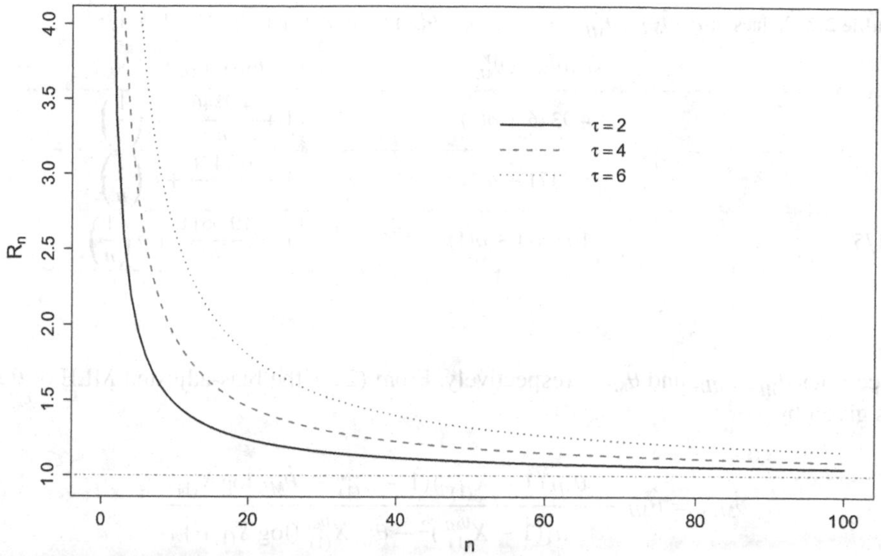

Fig. 2.1 Graph of the ratio $R_n(\hat{\theta}_{ML^*}, \hat{\theta}_{ML}^{\gamma})$ up to the order $1/n$ for $\tau = \theta - \gamma = 2, 4, 6$

$(i = 1, \dots, n)$, respectively, in Example 2.7.1, we have the same results including the UMVU estimation as those in Example 2.7.1. For Pareto distributions, see also Arnold (2015).

Example 2.7.4 **(Lower-truncated beta distribution)** Let $c = 0$, $d = 1$, $a(x) = x^{-1}$ and $u(x) = \log x$ for $0 < \gamma \le x < 1$ in the density (1.7). Since $b(\theta, \gamma) = \theta^{-1}(1-\gamma^{\theta})$ for $\theta \in \Theta = (0, \infty)$, it follows from (2.1) and Theorem 2.4.1 that

$$\lambda_1(\theta, \gamma) = -\frac{1}{\theta} - \frac{(\log \gamma)\gamma^{\theta}}{1-\gamma^{\theta}}, \quad \frac{\partial \lambda_1}{\partial \gamma}(\theta, \gamma) = -\frac{\gamma^{\theta-1}}{(1-\gamma^{\theta})^2}(1-\gamma^{\theta}+\theta \log \gamma),$$

$$\lambda_2(\theta, \gamma) = \frac{1}{\theta^2} - \frac{(\log \gamma)^2 \gamma^{\theta}}{(1-\gamma^{\theta})^2}, \quad k(\theta, \gamma) = \frac{\theta \gamma^{\theta-1}}{1-\gamma^{\theta}}.$$

Then, it follows from (2.2), (2.4), and (2.6) that the solution of θ of the following equations

$$\frac{1}{n}\sum_{i=1}^{n} \log X_i + \frac{1}{\theta} + \frac{(\log \gamma)\gamma^{\theta}}{1-\gamma^{\theta}} = 0,$$

$$\frac{1}{n}\sum_{i=1}^{n} \log X_i + \frac{1}{\theta} + \frac{(\log X_{(1)})X_{(1)}^{\theta}}{1-X_{(1)}^{\theta}} = 0,$$

$$\frac{1}{n-1}\sum_{i=2}^{n} \log X_{(i)} + \frac{1}{\theta} + \frac{(\log X_{(1)})X_{(1)}^{\theta}}{1-X_{(1)}^{\theta}} = 0$$

Table 2.2 Values of $d_n(\hat{\theta}_{ML^*}, \hat{\theta}^{\gamma}_{ML})$ and $R_n(\hat{\theta}_{ML^*}, \hat{\theta}^{\gamma}_{ML})$ for $\theta = 2$ and $\gamma = 1/2, 1/3, 1/5$

γ	$d_n(\hat{\theta}_{ML^*}, \hat{\theta}^{\gamma}_{ML})$	$R_n(\hat{\theta}_{ML^*}, \hat{\theta}^{\gamma}_{ML})$
1/2	$4.9346 + o(1)$	$1 + \dfrac{4.9346}{n} + o\left(\dfrac{1}{n}\right)$
1/3	$6.7471 + o(1)$	$1 + \dfrac{6.7471}{n} + o\left(\dfrac{1}{n}\right)$
1/5	$10.0611 + o(1)$	$1 + \dfrac{10.0611}{n} + o\left(\dfrac{1}{n}\right)$

becomes $\hat{\theta}^{\gamma}_{ML}$, $\hat{\theta}_{ML}$, and $\hat{\theta}_{MCL}$, respectively. From (2.5), the bias-adjusted MLE of θ is given by

$$\hat{\theta}_{ML^*} = \hat{\theta}_{ML} - \frac{\hat{\theta}_{ML}(1 - X_{(1)}^{\hat{\theta}_{ML}})(1 - X_{(1)}^{\hat{\theta}_{ML}} + \hat{\theta}_{ML} \log X_{(1)})}{n\{(1 - X_{(1)}^{\hat{\theta}_{ML}})^2 - \hat{\theta}^2_{ML} X_{(1)}^{\hat{\theta}_{ML}} (\log X_{(1)})^2\}}.$$

From Theorem 2.6.1, we obtain the second-order asymptotic losses

$$d_n(\hat{\theta}_{ML^*}, \hat{\theta}_{MCL}) = o(1),$$
$$d_n(\hat{\theta}_{ML^*}, \hat{\theta}^{\gamma}_{ML}) = d_n(\hat{\theta}_{MCL}, \hat{\theta}^{\gamma}_{ML}) = \frac{(1 - \gamma^{\theta} + \theta \log \gamma)^2}{(1 - \gamma^{\theta})^2 - \theta^2 \gamma^{\theta} (\log \gamma)^2}.$$

When $\theta = 2$ and $\gamma = 1/2, 1/3, 1/5$, the values of second-order asymptotic loss $d_n(\hat{\theta}_{ML^*}, \hat{\theta}^{\gamma}_{ML})$ and the ratio $R_n(\hat{\theta}_{ML^*}, \hat{\theta}^{\gamma}_{ML})$ up to the order $1/n$ are obtained from (2.8) and Remark 2.6.1 (see Table 2.2 and Fig. 2.2).

Example 2.7.5 (**Lower-truncated Erlang type distribution**) Let $c = 0, d = \infty$, $a(x) = |x|^{j-1}$ and $u(x) = -|x|$ for $-\infty < \gamma \le x < \infty$ in the density (1.7), where $j = 1, 2, \ldots$. Note that the distribution is a lower-truncated Erlang distribution when $\gamma > 0$ and a one-sided truncated bilateral exponential distribution when $j = 1$. Since for each $j = 1, 2, \ldots$,

$$b_j(\theta, \gamma) = \int_{\gamma}^{\infty} |x|^{j-1} e^{-\theta|x|} dx,$$

it follows that $\Theta = (0, \infty)$. Let j be arbitrarily fixed in $\{1, 2, \ldots\}$ and $\lambda_{ji}(\theta, \gamma) = (\partial^i/\partial\theta^i) \log b_j(\theta, \gamma)$ $(i = 1, 2, \ldots)$. Since $\partial b_j/\partial\theta = -b_{j+1}$, it follows from (2.1) and Theorem 2.4.1 that

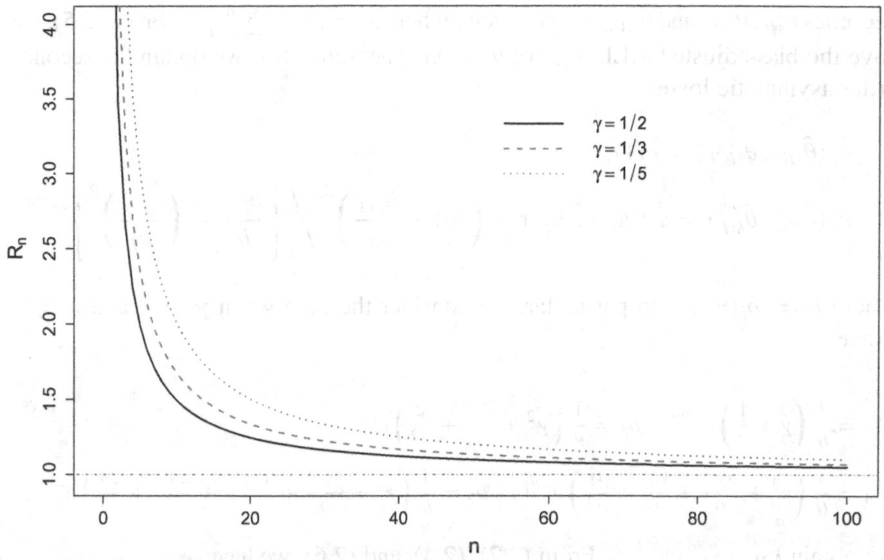

Fig. 2.2 Graph of the ratio $R_n(\hat{\theta}_{ML^*}, \hat{\theta}^{\gamma}_{ML})$ up to the order $1/n$ for $\theta = 2$ and $\gamma = 1/2, 1/3, 1/5$

$$\lambda_{j1}(\theta, \gamma) = -\frac{b_{j+1}(\theta, \gamma)}{b_j(\theta, \gamma)},$$

$$\frac{\partial \lambda_{j1}}{\partial \gamma}(\theta, \gamma) = \begin{cases} \frac{\gamma^{j-1} e^{-\theta\gamma}}{b_j(\theta,\gamma)} \left\{ \frac{b_{j+1}(\theta,\gamma)}{b_j(\theta,\gamma)} + \gamma \right\} & \text{for } \gamma > 0, \\ (-1)^j \frac{\gamma^{j-1} e^{\theta\gamma}}{b_j(\theta,\gamma)} \left\{ \frac{b_{j+1}(\theta,\gamma)}{b_j(\theta,\gamma)} + \gamma \right\} & \text{for } \gamma \le 0, \end{cases}$$

$$\lambda_{j2}(\theta, \gamma) = \frac{b_{j+2}(\theta, \gamma)}{b_j(\theta, \gamma)} - \left\{ \frac{b_{j+1}(\theta, \gamma)}{b_j(\theta, \gamma)} \right\}^2,$$

$$\lambda_{j3}(\theta, \gamma) = -\frac{b_{j+3}(\theta, \gamma)}{b_j(\theta, \gamma)} + \frac{3b_{j+1}(\theta, \gamma) b_{j+2}(\theta, \gamma)}{b_j^2(\theta, \gamma)} - 2\left\{ \frac{b_{j+1}(\theta, \gamma)}{b_j(\theta, \gamma)} \right\}^3,$$

$$k_j(\theta, \gamma) = \frac{|\gamma|^{j-1} e^{-\theta|\gamma|}}{b_j(\theta, \gamma)}.$$

Then, it follows from (2.2), (2.4), and (2.6) that the solutions of θ of the equations

$$\bar{X} - \frac{b_{j+1}(\theta, \gamma)}{b_j(\theta, \gamma)} = 0, \quad \bar{X} - \frac{b_{j+1}(\hat{\theta}_{ML}, X_{(1)})}{b_j(\hat{\theta}_{ML}, X_{(1)})} = 0,$$

$$\frac{1}{n-1} \sum_{i=2}^{n} X_{(i)} - \frac{b_{j+1}(\theta, X_{(1)})}{b_j(\theta, X_{(1)})} = 0$$

becomes $\hat{\theta}_{ML}^{\gamma}$, $\hat{\theta}_{ML}$, and $\hat{\theta}_{MCL}$, respectively, where $\bar{X} = (1/n)\sum_{i=1}^{n} X_i$. From (2.5), we have the bias-adjusted MLE $\hat{\theta}_{ML^*}$ of θ. From Theorem 2.6.1, we obtain the second-order asymptotic losses

$$d_n(\hat{\theta}_{ML^*}, \hat{\theta}_{MCL}) = o(1),$$

$$d_n(\hat{\theta}_{ML^*}, \hat{\theta}_{ML}^{\gamma}) = d_n(\hat{\theta}_{MCL}, \hat{\theta}_{ML}^{\gamma}) = \left(|\gamma| - \frac{b_{j+1}}{b_j}\right)^2 \bigg/ \left\{\frac{b_{j+2}}{b_j} - \left(\frac{b_{j+1}}{b_j}\right)^2\right\},$$

where $b_j = b_j(\theta, \gamma)$. In particular, we consider the case when $\gamma \geq 0$ and $j = 2$. Since

$$b_2 = \frac{1}{\theta}\left(\gamma + \frac{1}{\theta}\right)e^{-\theta\gamma}, \quad b_3 = \frac{1}{\theta}\left(\gamma^2 + \frac{2\gamma}{\theta} + \frac{2}{\theta^2}\right)e^{-\theta\gamma},$$

$$b_4 = \frac{1}{\theta}\left(\gamma^3 + \frac{3\gamma^2}{\theta} + \frac{6\gamma}{\theta^2} + \frac{6}{\theta^3}\right)e^{-\theta\gamma}, \quad b_5 = \frac{1}{\theta}\left(\gamma^4 + \frac{4\gamma^3}{\theta} + \frac{12\gamma^2}{\theta^2} + \frac{24\gamma}{\theta^3} + \frac{24}{\theta^4}\right)e^{-\theta\gamma},$$

we obtain λ_{21}, λ_{22}, and λ_{23}. From (2.2), (2.4), and (2.6), we have

$$\hat{\theta}_{ML}^{\gamma} = 4\left\{\bar{X} - 2\gamma + \sqrt{4\gamma(\bar{X} - \gamma) + \bar{X}^2}\right\}^{-1},$$

$$\hat{\theta}_{ML} = 4\left\{\bar{X} - 2X_{(1)} + \sqrt{4X_{(1)}(\bar{X} - X_{(1)}) + \bar{X}^2}\right\}^{-1},$$

$$\hat{\theta}_{MCL} = 4\left\{\tilde{X} - 2X_{(1)} + \sqrt{4X_{(1)}(\tilde{X} - X_{(1)}) + \tilde{X}^2}\right\}^{-1},$$

where $\tilde{X} = (1/(n-1))\sum_{i=2}^{n} X_{(i)}$. From (2.5), we also obtain the bias-adjusted MLE $\hat{\theta}_{ML^*}$ of θ. Further, we have the second-order asymptotic loss

$$d_n(\hat{\theta}_{ML^*}, \hat{\theta}_{ML}^{\gamma}) = d_n(\hat{\theta}_{MCL}, \hat{\theta}_{ML}^{\gamma}) = \frac{(\theta\gamma + 2)^2}{(\theta\gamma + 2)^2 - 2} + o(1)$$

and the ratio

$$R_n(\hat{\theta}_{ML^*}, \hat{\theta}_{ML}^{\gamma}) = R_n(\hat{\theta}_{MCL}, \hat{\theta}_{ML}^{\gamma}) = 1 + \frac{(\theta\gamma + 2)^2}{n\{(\theta\gamma + 2)^2 - 2\}} + O\left(\frac{1}{n}\right).$$

If $\gamma = 0$, then $\hat{\theta}_{ML}^{0} = 2/\bar{X}$ and

$$d_n(\hat{\theta}_{ML^*}, \hat{\theta}_{ML}^{0}) = d_n(\hat{\theta}_{MCL}, \hat{\theta}_{ML}^{0}) = 2 + o(1),$$

$$R_n(\hat{\theta}_{ML^*}, \hat{\theta}_{ML}^{0}) = R_n(\hat{\theta}_{MCL}, \hat{\theta}_{ML}^{0}) = 1 + \frac{2}{n} + O\left(\frac{1}{n}\right).$$

In Vancak et al. (2015), the ratio of the mean squared error (MSE) of $\hat{\theta}_{ML}$ to that of $\hat{\theta}^0_{ML}$ is calculated by simulation and its graph is given as a function of n when $\theta = -1$. Here, we can theoretically obtain the function. Indeed, letting $\gamma = 0$ and

$$U_0 = \sqrt{\lambda_{22}(\theta, 0)}(\hat{\theta}^0_{ML} - \theta),$$

we have from (2.13) in Appendix A1 later

$$E_\theta(U_0^2) = 1 + \frac{1}{n}\left\{\frac{11\lambda_{23}^2(\theta, 0)}{4\lambda_{22}^3(\theta, 0)} - \frac{\lambda_{24}(\theta, 0)}{\lambda_{22}^2(\theta, 0)}\right\} + O\left(\frac{1}{n\sqrt{n}}\right),$$

hence the MSE of $\hat{\theta}^0_{ML}$ is given by

$$\begin{aligned} MSE_\theta(\hat{\theta}^0_{ML}) &= E_\theta[(\hat{\theta}^0_{ML} - \theta)^2] = \frac{1}{\lambda_{22}(\theta, 0)n}E_\theta(U_0^2) \\ &= \frac{1}{\lambda_{22}(\theta, 0)n}\left[1 + \frac{1}{n}\left\{\frac{11\lambda_{23}^2(\theta, 0)}{4\lambda_{22}^3(\theta, 0)} - \frac{\lambda_{24}(\theta, 0)}{\lambda_{22}^2(\theta, 0)}\right\} + O\left(\frac{1}{n\sqrt{n}}\right)\right]. \end{aligned}$$

When γ is unknown, letting

$$\hat{U} = \sqrt{\lambda_{22}(\theta, \gamma)}(\hat{\theta}_{ML} - \theta),$$

we have from (2.24) in Appendix A1 given later

$$\begin{aligned} E_\theta(\hat{U}^2) =& 1 - \frac{2}{k_2\lambda_{22}n}\left(\frac{\partial\lambda_{21}}{\partial\gamma}\right)\left\{u(\gamma) - \lambda_{21} + \frac{1}{k_2}\left(\frac{\partial\lambda_{21}}{\partial\gamma}\right)\right\} + \frac{11\lambda_{23}^2}{4\lambda_{22}^3 n} \\ &+ \frac{3\lambda_{23}}{k_2\lambda_{22}^2 n}\left(\frac{\partial\lambda_{21}}{\partial\gamma}\right) - \frac{2}{k_2\lambda_{22}n}\left(\frac{\partial\lambda_{22}}{\partial\gamma}\right) - \frac{\lambda_{24}}{\lambda_{22}^2 n} + O\left(\frac{1}{n\sqrt{n}}\right), \end{aligned}$$

where $k_2 = k_2(\theta, \gamma) = a(\gamma)e^{\theta u(\gamma)}/b_2(\theta, \gamma)$ and $\lambda_{2j} = \lambda_{2j}(\theta, \gamma)$ $(j = 1, 2, 3, 4)$. From (2.28), and (2.29) in Appendix A1 given later, we have

$$\frac{\partial\lambda_{21}}{\partial\gamma}(\theta, \gamma) = k_2(\theta, \gamma)\{\lambda_{21}(\theta, \gamma) - u(\gamma)\}, \qquad \frac{\partial k_2}{\partial\theta}(\theta, \gamma) = k_2(\theta, \gamma)\{u(\gamma) - \lambda_{21}(\theta, \gamma)\}.$$

Since

$$\frac{\partial b_2}{\partial\gamma}(\theta, \gamma) = -k_2(\theta, \gamma)b_2(\theta, \gamma),$$

it follows that

$$\frac{\partial\lambda_{22}}{\partial\gamma}(\theta, \gamma) = -\frac{\partial^2 k_2}{\partial\theta^2}(\theta, \gamma) = -k_2(\theta, \gamma)\{u(\gamma) - \lambda_{21}\}^2 + k_2(\theta, \gamma)\lambda_{22}(\theta, \gamma),$$

hence

$$E_{\theta,\gamma}(\hat{U}^2) = 1 + \frac{11\lambda_{23}^2}{4\lambda_{22}^3 n} - \frac{\lambda_{24}}{\lambda_{22}^2 n} - \frac{3\lambda_{23}}{\lambda_{22}^2 n}(u(\gamma) - \lambda_{21}) + \frac{2}{\lambda_{22} n}(u(\gamma) - \lambda_{21})^2 - \frac{2}{n} + O\left(\frac{1}{n\sqrt{n}}\right).$$

Denote the ratio of the MSE of $\hat{\theta}_{ML}$ at $\gamma = 0$ to that of $\hat{\theta}_{ML}^0$ by

$$R_{MSE}(\hat{\theta}_{ML}, \hat{\theta}_{ML}^0) := [MSE_{\theta,\gamma}(\hat{\theta}_{ML})]_{\gamma=0} \Big/ MSE_{\theta}(\hat{\theta}_{ML}^0).$$

Since

$$MSE_{\theta,\gamma}(\hat{\theta}_{ML}) = E_{\theta,\gamma}[(\hat{\theta}_{ML} - \theta)^2] = \frac{1}{\lambda_{22} n} E_{\theta,\gamma}(\hat{U}^2),$$

we have

$$[MSE_{\theta,\gamma}(\hat{\theta}_{ML})]_{\gamma=0} = \frac{1}{\lambda_{22}(\theta,0)n}\left\{1 + \frac{11\lambda_{23}^2(\theta,0)}{4\lambda_{22}^3(\theta,0)n} - \frac{\lambda_{24}(\theta,0)}{\lambda_{22}^2(\theta,0)n} - \frac{3\lambda_{21}(\theta,0)\lambda_{23}(\theta,0)}{\lambda_{22}^2(\theta,0)n} + \frac{2\lambda_{21}^2(\theta,0)}{\lambda_{22}(\theta,0)n} - \frac{2}{n} + O\left(\frac{1}{n\sqrt{n}}\right)\right\}.$$

Hence, we obtain

$$\begin{aligned} R_{MSE}(\hat{\theta}_{ML}, \hat{\theta}_{ML}^0) =& \left\{1 + \frac{11\lambda_{23}^2(\theta,0)}{4\lambda_{22}^3(\theta,0)n} - \frac{\lambda_{24}(\theta,0)}{\lambda_{22}^2(\theta,0)n} + \frac{3\lambda_{21}(\theta,0)\lambda_{23}(\theta,0)}{\lambda_{22}^2(\theta,0)n} + \frac{2\lambda_{21}^2(\theta,0)}{\lambda_{22}(\theta,0)n} \right. \\ & \left. - \frac{2}{n} + O\left(\frac{1}{n\sqrt{n}}\right)\right\} \cdot \left[1 - \frac{1}{n}\left\{\frac{11\lambda_{23}^2(\theta,0)}{4\lambda_{22}^3(\theta,0)} - \frac{\lambda_{24}(\theta,0)}{\lambda_{22}^2(\theta,0)}\right\} + O\left(\frac{1}{n\sqrt{n}}\right)\right] \\ =& 1 + \frac{1}{n}\left\{\frac{3\lambda_{21}(\theta,0)\lambda_{23}(\theta,0)}{\lambda_{22}^2(\theta,0)} + \frac{2\lambda_{21}^2(\theta,0)}{\lambda_{22}(\theta,0)n} - 2\right\} + O\left(\frac{1}{n\sqrt{n}}\right). \end{aligned}$$

Since

$$\lambda_{21}(\theta,0) = -\frac{b_3(\theta,0)}{b_2(\theta,0)} = -\frac{2}{\theta},$$

$$\lambda_{22}(\theta,0) = \frac{b_4(\theta,0)}{b_2(\theta,0)} - \left\{\frac{b_3(\theta,0)}{b_2(\theta,0)}\right\}^2 = \frac{2}{\theta^2},$$

$$\lambda_{23}(\theta,0) = -\frac{b_5(\theta,0)}{b_2(\theta,0)} + \frac{3b_3(\theta,0)b_4(\theta,0)}{b_2^2(\theta,0)} - 2\left\{\frac{b_3(\theta,0)}{b_2(\theta,0)}\right\}^2 = -\frac{4}{\theta^3},$$

it follows that

$$R_{MSE}(\hat{\theta}_{ML}, \hat{\theta}^0_{ML}) = 1 + \frac{8}{n} + O\left(\frac{1}{n\sqrt{n}}\right),$$

which is the required function of n. The ratio $R_{MSE}(\hat{\theta}_{ML}, \hat{\theta}^0_{ML})$ seems to be fit for the simulation result, i.e., Fig. 3 by Vancak et al. (2015).

Example 2.7.6 (**Lower-truncated lognormal distribution**) Let $c = 0, d = \infty$, $a(x) = x^{-1}\exp\{-(1/2)(\log x)^2\}$ and $u(x) = \log x$ for $0 < \gamma \leq x < \infty$ in the density (1.7). Then, $b(\theta, \gamma) = \Phi(\theta - \log\gamma)/\phi(\theta)$ for $\theta \in \Theta = (-\infty, \infty)$, where $\Phi(x) = \int_{-\infty}^{x}\phi(t)dt$ with $\phi(t) = (1/\sqrt{2\pi})e^{-t^2/2}$ for $-\infty < t < \infty$. Letting $t = \log x$ and $\gamma_0 = \log\gamma$, we see that (1.7) becomes

$$f(t;\theta,\gamma_0) = \begin{cases} \frac{1}{\sqrt{2\pi}\Phi(\theta-\gamma_0)}e^{-(t-\theta)^2/2} & \text{for } -\infty < \gamma_0 \leq \text{t} < \infty, \\ 0 & \text{otherwise.} \end{cases}$$

Hence, the lower-truncated lognormal case is reduced to the truncated normal one in Example 2.7.2.

For a truncated beta distribution and a truncated Erlang distribution, related results to the above can be found in Vancak et al. (2015).

2.8 Concluding Remarks

In a oTEF of distributions with a two-dimensional parameter (θ, γ), we considered the estimation problem of a natural parameter θ in the presence of a truncation parameter γ as a nuisance parameter. In the paper of Bar-Lev (1984), it was shown that the MLE $\hat{\theta}^{\gamma}_{ML}$ of θ for known γ, the MLE $\hat{\theta}_{ML}$ and the MCLE $\hat{\theta}_{MCL}$ of θ for unknown γ were asymptotically equivalent in the sense that they had the same asymptotic normal distribution. In this chapter, we derived the stochastic expansions of $\hat{\theta}^{\gamma}_{ML}$, $\hat{\theta}_{ML}$, and $\hat{\theta}_{MCL}$. We also obtained the second-order asymptotic loss of the bias-adjusted MLE $\hat{\theta}_{ML^*}$ relative to $\hat{\theta}^{\gamma}_{ML}$ from their second-order asymptotic variances and showed that $\hat{\theta}_{ML^*}$ and $\hat{\theta}_{MCL}$ were second order asymptotically equivalent in the sense that their asymptotic variances were same up to the second order, i.e., $o(1/n)$ as in (2.7). It seems to be natural that $\hat{\theta}^{\gamma}_{ML}$ is second order asymptotically better than $\hat{\theta}_{ML^*}$ after adjusting the bias of $\hat{\theta}_{ML}$ such that $\hat{\theta}_{ML}$ has the same as that of $\hat{\theta}^{\gamma}_{ML}$. The values of the second-order asymptotic losses of $\hat{\theta}_{ML^*}$ and $\hat{\theta}_{MCL}$ given by (2.8) and (2.9) are quite simple, which results from the truncated exponential family $\mathscr{P}_o$ of distributions.

The corresponding results to Theorems 2.3.1, 2.4.1, 2.5.1, and 2.6.1 can be obtained in the case of a two-sided truncated exponential family of distributions with a natural parameter θ and two truncation parameters γ and ν as nuisance parameters, including an upper-truncated Pareto distribution which is important in applications (see Chap. 3). Further, they may be similarly extended to the case of a more general truncated family of distributions from the truncated exponential family $\mathscr{P}_o$. In relation to Theorem 2.4.1, if two different bias-adjustments are introduced, i.e., $\hat{\theta}_{ML} + (1/n)c_i(\hat{\theta}_{ML})$ $(i = 1, 2)$, then the problem whether or not the admissibility result holds may be interesting.

2.9 Appendix A1

The proof of Theorem 2.3.1 Let $\lambda_i = \lambda_i(\theta, \gamma)$ $(i = 1, 2, 3, 4)$. Since

$$Z_1 = \frac{1}{\sqrt{\lambda_2 n}} \sum_{i=1}^{n} \{u(X_i) - \lambda_1\}, \quad U_\gamma := \sqrt{\lambda_2 n}(\hat{\theta}_{ML}^{\gamma} - \theta),$$

by the Taylor expansion, we obtain from (2.2)

$$0 = \sqrt{\frac{\lambda_2}{n}} Z_1 - \sqrt{\frac{\lambda_2}{n}} U_\gamma - \frac{\lambda_3}{2\lambda_2 n} U_\gamma^2 - \frac{\lambda_4}{6\lambda_2^{3/2} n\sqrt{n}} U_\gamma^3 + O_p\left(\frac{1}{n^2}\right),$$

which implies that the stochastic expansion of U_γ is given by

$$U_\gamma = Z_1 - \frac{\lambda_3}{2\lambda_2^{3/2}\sqrt{n}} Z_1^2 + \frac{1}{2n}\left(\frac{\lambda_3^2}{\lambda_2^3} - \frac{\lambda_4}{3\lambda_2^2}\right) Z_1^3 + O_p\left(\frac{1}{n\sqrt{n}}\right). \tag{2.10}$$

Since

$$E_\theta(Z_1) = 0, \; V_\theta(Z_1) = E_\theta(Z_1^2) = 1,$$
$$E_\theta(Z_1^3) = \frac{\lambda_3}{\lambda_2^{3/2}\sqrt{n}}, \quad E_\theta(Z_1^4) = 3 + \frac{\lambda_4}{\lambda_2^2 n}, \tag{2.11}$$

it follows that

$$E_\theta(U_\gamma) = -\frac{\lambda_3}{2\lambda_2^{3/2}\sqrt{n}} + O\left(\frac{1}{n\sqrt{n}}\right), \tag{2.12}$$

$$E_\theta(U_\gamma^2) = 1 + \frac{1}{n}\left(\frac{11\lambda_3^2}{4\lambda_2^3} - \frac{\lambda_4}{\lambda_2^2}\right) + O\left(\frac{1}{n\sqrt{n}}\right), \tag{2.13}$$

hence, by (2.12) and (2.13)

$$V_\theta(U_\gamma) = 1 + \frac{1}{n}\left(\frac{5\lambda_3^2}{2\lambda_2^3} - \frac{\lambda_4}{\lambda_2^2}\right) + o\left(\frac{1}{n\sqrt{n}}\right). \tag{2.14}$$

From (2.10), (2.12), and (2.14), we have the conclusion of Theorem 2.3.1.

Before proving Theorem 2.4.1, we prepare three lemmas (the proofs are given in Appendix A2).

Lemma 2.9.1 *The second-order asymptotic density of $T_{(1)}$ is given by*

$$\begin{aligned} f_{T_{(1)}}(t) =& k(\theta,\gamma)e^{-k(\theta,\gamma)t} \\ & - \frac{1}{2n}\left\{\frac{\partial}{\partial\gamma}\log k(\theta,\gamma)\right\}\left\{k(\theta,\gamma)t^2 - 2t\right\}k(\theta,\gamma)e^{-k(\theta,\gamma)t} + O\left(\frac{1}{n^2}\right) \end{aligned} \tag{2.15}$$

for $t > 0$, where $k(\theta,\gamma) := a(\gamma)e^{\theta u(\gamma)}/b(\theta,\gamma)$ and

$$E_{\theta,\gamma}(T_{(1)}) = \frac{1}{k(\theta,\gamma)} + \frac{A(\theta,\gamma)}{n} + O\left(\frac{1}{n^2}\right),\quad E_{\theta,\gamma}(T_{(1)}^2) = \frac{2}{k^2(\theta,\gamma)} + O\left(\frac{1}{n}\right), \tag{2.16}$$

where

$$A(\theta,\gamma) := -\frac{1}{k^2(\theta,\gamma)}\left\{\frac{\partial}{\partial\gamma}\log k(\theta,\gamma)\right\}.$$

Lemma 2.9.2 *It holds that*

$$E_{\theta,\gamma}(Z_1 T_{(1)}) = \frac{1}{k\sqrt{\lambda_2 n}}\left\{u(\gamma) - \lambda_1 + \frac{2}{k}\left(\frac{\partial\lambda_1}{\partial\gamma}\right)\right\} + O\left(\frac{1}{n\sqrt{n}}\right), \tag{2.17}$$

where $k = k(\theta,\gamma)$ and $\lambda_i = \lambda_i(\theta,\gamma)$ $(i = 1, 2)$.

Lemma 2.9.3 *It holds that*

$$E_{\theta,\gamma}(Z_1^2 T_{(1)}) = \frac{1}{k} + O\left(\frac{1}{n}\right), \tag{2.18}$$

where $k = k(\theta,\gamma)$.

The proof of Theorem 2.4.1 Since, for $(\theta, \gamma) \in \Theta \times (c, X_{(1)})$

$$
\begin{aligned}
&\lambda_1(\hat{\theta}_{ML}, X_{(1)}) \\
&= \lambda_1(\theta, \gamma) + \left\{ \frac{\partial}{\partial \theta} \lambda_1(\theta, \gamma) \right\} (\hat{\theta}_{ML} - \theta) + \left\{ \frac{\partial}{\partial \gamma} \lambda_1(\theta, \gamma) \right\} (X_{(1)} - \gamma) \\
&+ \frac{1}{2} \left\{ \frac{\partial^2}{\partial \theta^2} \lambda_1(\theta, \gamma) \right\} (\hat{\theta}_{ML} - \theta)^2 + \left\{ \frac{\partial^2}{\partial \theta \partial \gamma} \lambda_1(\theta, \gamma) \right\} (\hat{\theta}_{ML} - \theta)(X_{(1)} - \gamma) \\
&+ \frac{1}{2} \left\{ \frac{\partial^2}{\partial \gamma^2} \lambda_1(\theta, \gamma) \right\} (X_{(1)} - \gamma)^2 + \frac{1}{6} \left\{ \frac{\partial^3}{\partial \theta^3} \lambda_1(\theta, \gamma) \right\} (\hat{\theta}_{ML} - \theta)^3 \\
&+ \frac{1}{2} \left\{ \frac{\partial^2}{\partial \theta^2} \lambda_1(\theta, \gamma) \right\} \left\{ \frac{\partial}{\partial \gamma} \lambda_1(\theta, \gamma) \right\} (\hat{\theta}_{ML} - \theta)^2 (X_{(1)} - \gamma) + \cdots, \qquad (2.19)
\end{aligned}
$$

noting $\hat{U} = \sqrt{\lambda_2 n}(\hat{\theta}_{ML} - \theta)$ and $T_{(1)} = n(X_{(1)} - \gamma)$, we have from (2.4) and (2.19)

$$
\begin{aligned}
0 = &\sqrt{\frac{\lambda_2}{n}} Z_1 - \sqrt{\frac{\lambda_2}{n}} \hat{U} - \frac{1}{n} \left(\frac{\partial \lambda_1}{\partial \gamma} \right) T_{(1)} - \frac{\lambda_3}{2\lambda_2 n} \hat{U}^2 - \frac{1}{\sqrt{\lambda_2 n} n} \left(\frac{\partial \lambda_2}{\partial \gamma} \right) \hat{U} T_{(1)} \\
&- \frac{\lambda_4}{6\lambda_2^{3/2} n\sqrt{n}} \hat{U}^3 + O_p\left(\frac{1}{n^2} \right),
\end{aligned}
$$

where $\lambda_j = \lambda_j(\theta, \gamma)$ $(j = 1, 2, 3, 4)$ are defined by (2.1), hence the stochastic expansion of $\hat{U}$ is given by

$$
\begin{aligned}
\hat{U} = &Z_1 - \frac{1}{\sqrt{\lambda_2 n}} \left(\frac{\partial \lambda_1}{\partial \gamma} \right) T_{(1)} - \frac{\lambda_3}{2\lambda_2^{3/2} \sqrt{n}} Z_1^2 + \frac{\delta}{\lambda_2 n} Z_1 T_{(1)} \\
&+ \frac{1}{2n} \left(\frac{\lambda_3^2}{\lambda_2^3} - \frac{\lambda_4}{3\lambda_2^2} \right) Z_1^3 + O_p\left(\frac{1}{n\sqrt{n}} \right). \qquad (2.20)
\end{aligned}
$$

It follows from (2.11) and (2.20) that

$$
E_{\theta,\gamma}(\hat{U}) = -\frac{1}{\sqrt{\lambda_2 n}} \left(\frac{\partial \lambda_1}{\partial \gamma} \right) E_{\theta,\gamma}(T_{(1)}) - \frac{\lambda_3}{2\lambda_2^{3/2} \sqrt{n}} + \frac{\delta}{\lambda_2 n} E_{\theta,\gamma}(Z_1 T_{(1)}) + O\left(\frac{1}{n\sqrt{n}} \right). \qquad (2.21)
$$

Substituting (2.16) and (2.17) into (2.21), we obtain

$$
E_{\theta,\gamma}(\hat{U}) = -\frac{1}{\sqrt{\lambda_2 n}} \left\{ \frac{1}{k} \left(\frac{\partial \lambda_1}{\partial \gamma} \right) + \frac{\lambda_3}{2\lambda_2} \right\} + O\left(\frac{1}{n\sqrt{n}} \right), \qquad (2.22)
$$

where $k = k(\theta, \gamma)$ is defined in Lemma 2.9.1. We have from (2.20)

$$\begin{aligned} E_{\theta,\gamma}(\hat{U}^2) &= E_{\theta,\gamma}(Z_1^2) - \frac{1}{\sqrt{\lambda_2 n}}\left\{2\left(\frac{\partial \lambda_1}{\partial \gamma}\right) E_{\theta,\gamma}(Z_1 T_{(1)}) + \frac{\lambda_3}{\lambda_2} E_{\theta,\gamma}(Z_1^3)\right\} \\ &+ \frac{1}{\lambda_2 n}\left(\frac{\partial \lambda_1}{\partial \gamma}\right)^2 E_{\theta,\gamma}\left(T_{(1)}^2\right) + \frac{1}{\lambda_2 n}\left\{\frac{\lambda_3}{\lambda_2}\left(\frac{\partial \lambda_1}{\partial \gamma}\right) + 2\delta\right\} E_{\theta,\gamma}(Z_1^2 T_{(1)}) \\ &+ \frac{1}{n}\left(\frac{5\lambda_3^2}{4\lambda_2^3} - \frac{\lambda_4}{3\lambda_2^2}\right) E_{\theta,\gamma}(Z_1^4) + O\left(\frac{1}{n\sqrt{n}}\right). \end{aligned} \tag{2.23}$$

Substituting (2.11) and (2.16)–(2.18) into (2.23), we have

$$\begin{aligned} E_{\theta,\gamma}(\hat{U}^2) = 1 &- \frac{2}{k\lambda_2 n}\left(\frac{\partial \lambda_1}{\partial \gamma}\right)\left\{u(\gamma) - \lambda_1 + \frac{1}{k}\left(\frac{\partial \lambda_1}{\partial \gamma}\right)\right\} + \frac{11\lambda_3^2}{4\lambda_2^3 n} \\ &+ \frac{3\lambda_3}{k\lambda_2^2 n}\left(\frac{\partial \lambda_1}{\partial \gamma}\right) - \frac{2}{k\lambda_2 n}\left(\frac{\partial \lambda_2}{\partial \gamma}\right) - \frac{\lambda_4}{\lambda_2^2 n} + O\left(\frac{1}{n\sqrt{n}}\right). \end{aligned} \tag{2.24}$$

Since

$$\begin{aligned} &\frac{\sqrt{\lambda_2}\{(\partial/\partial\gamma)\lambda_1(\hat{\theta}_{ML}, X_{(1)})\}}{k(\hat{\theta}_{ML}, X_{(1)})\lambda_2(\hat{\theta}_{ML}, X_{(1)})\sqrt{n}} \\ &= \frac{(\partial/\partial\gamma)\lambda_1(\theta,\gamma)}{k\sqrt{\lambda_2 n}} + \frac{1}{k\lambda_2 n}\left\{\frac{\partial \lambda_2}{\partial \gamma}(\theta,\gamma) - \left(\frac{\lambda_3}{\lambda_2} + \frac{1}{k}\frac{\partial k}{\partial \theta}\right)\left(\frac{\partial \lambda_1}{\partial \gamma}\right)\right\}\hat{U} + O_p\left(\frac{1}{n\sqrt{n}}\right), \end{aligned}$$

it follows from (2.5) that the stochastic expansion of $\hat{U}^*$ is given by

$$\begin{aligned} \hat{U}^* &:= \sqrt{\lambda_2 n}(\hat{\theta}_{ML^*} - \theta) = \sqrt{\lambda_2 n}(\hat{\theta}_{ML} - \theta) + \frac{\sqrt{\lambda_2}\{(\partial/\partial\gamma)\lambda_1(\hat{\theta}_{ML}, X_{(1)})\}}{k(\hat{\theta}_{ML}, X_{(1)})\hat{\lambda}_2\sqrt{n}} \\ &= \hat{U} + \frac{1}{k\sqrt{\lambda_2 n}}\left(\frac{\partial \lambda_1}{\partial \gamma}\right) - \frac{1}{k\lambda_2 n}\left\{\delta + \frac{1}{k}\left(\frac{\partial k}{\partial \theta}\right)\left(\frac{\partial \lambda_1}{\partial \gamma}\right)\right\} Z_1 + O_p\left(\frac{1}{n\sqrt{n}}\right), \end{aligned} \tag{2.25}$$

where $\hat{U}$ is given by (2.20), $\lambda_i = \lambda_i(\theta, \gamma)$ $(i = 1, 2, 3)$ and $k = k(\theta, \gamma)$. From (2.11) and (2.22), we have

$$E_{\theta,\gamma}(\hat{U}^*) = -\frac{\lambda_3}{2\lambda_2^{3/2}\sqrt{n}} + O\left(\frac{1}{n\sqrt{n}}\right). \tag{2.26}$$

It follows from (2.22), (2.24), and (2.25) that

$$E_{\theta,\gamma}(\hat{U}^{*^2}) = 1 - \frac{2}{k\lambda_2 n}\left(\frac{\partial\lambda_1}{\partial\gamma}\right)\left\{u(\gamma) - \lambda_1 + \frac{3}{2k}\left(\frac{\partial\lambda_1}{\partial\gamma}\right)\right\} + \frac{11\lambda_3^2}{4\lambda_2^3 n} - \frac{\lambda_4}{\lambda_2^2 n} - \frac{2}{k^2\lambda_2 n}\left(\frac{\partial\lambda_1}{\partial\gamma}\right)\left(\frac{\partial k}{\partial\theta}\right) + O\left(\frac{1}{n\sqrt{n}}\right),$$

hence, by (2.26)

$$V_{\theta,\gamma}(\hat{U}^*) = 1 + \frac{1}{n}\left(\frac{5\lambda_3^2}{2\lambda_2^3} - \frac{\lambda_4}{\lambda_2^2}\right) - \frac{2}{k\lambda_2 n}\left(\frac{\partial\lambda_1}{\partial\gamma}\right)\left\{u(\gamma) - \lambda_1 + \frac{1}{k}\left(\frac{\partial k}{\partial\theta}\right)\right\} - \frac{3}{k^2\lambda_2 n}\left(\frac{\partial\lambda_1}{\partial\gamma}\right)^2 + O\left(\frac{1}{n\sqrt{n}}\right). \tag{2.27}$$

Since, by (2.1)

$$\lambda_1(\theta,\gamma) = \frac{\partial}{\partial\theta}\log b(\theta,\gamma) = \frac{1}{b(\theta,\gamma)}\int_\gamma^d a(x)u(x)e^{\theta u(x)}dx,$$

it follows that

$$\frac{\partial\lambda_1(\theta,\gamma)}{\partial\gamma} = \frac{a(\gamma)e^{\theta u(\gamma)}}{b(\theta,\gamma)}\{\lambda_1(\theta,\gamma) - u(\gamma)\} = k(\theta,\gamma)\{\lambda_1(\theta,\gamma) - u(\gamma)\}. \tag{2.28}$$

Since

$$\frac{\partial k}{\partial\theta}(\theta,\gamma) = k(\theta,\gamma)\{u(\gamma) - \lambda_1(\theta,\gamma)\}, \tag{2.29}$$

it is seen from (2.27)–(2.29) that

$$V_{\theta,\gamma}(\hat{U}^*) = 1 + \frac{1}{n}\left(\frac{5\lambda_3^2}{2\lambda_2^3} - \frac{\lambda_4}{\lambda_2^2}\right) + \frac{1}{\lambda_2 n}\{\lambda_1 - u(\gamma)\}^2 + O\left(\frac{1}{n\sqrt{n}}\right). \tag{2.30}$$

From (2.25), (2.26) and (2.30), we have the conclusion of Theorem 2.4.1.

The proof of Theorem 2.5.1 Since, from (2.6)

$$0 = \frac{1}{n-1}\sum_{i=2}^n\{u(Y_i) - \lambda_1(\theta, x_{(1)})\} - \frac{1}{\sqrt{n}}\lambda_2(\theta, x_{(1)})\sqrt{n}(\hat{\theta}_{MCL} - \theta) - \frac{1}{2n}\lambda_3(\theta, x_{(1)})n(\hat{\theta}_{MCL} - \theta)^2 - \frac{1}{6n\sqrt{n}}\lambda_4(\theta, x_{(1)})n\sqrt{n}(\hat{\theta}_{MCL} - \theta)^3 + O_p\left(\frac{1}{n^2}\right),$$

letting

$$\tilde{Z}_1 = \frac{1}{\sqrt{\tilde{\lambda}_2(n-1)}}\sum_{i=2}^{n}\{u(Y_i) - \lambda_1(\theta, x_{(1)})\}, \quad \tilde{U} = \sqrt{\tilde{\lambda}_2 n}(\hat{\theta}_{MCL} - \theta),$$

where $\tilde{\lambda}_i := \lambda_i(\theta, x_{(1)})$ $(i = 1, 2, 3, 4)$, we have

$$0 = \sqrt{\frac{\tilde{\lambda}_2}{n-1}}\tilde{Z}_1 - \sqrt{\frac{\tilde{\lambda}_2}{n}}\tilde{U} - \frac{\tilde{\lambda}_3}{2\tilde{\lambda}_2 n}\tilde{U}^2 - \frac{\tilde{\lambda}_4}{6\tilde{\lambda}_2^{3/2} n\sqrt{n}}\tilde{U}^3 + O_p\left(\frac{1}{n^2}\right),$$

hence, the stochastic expansion of $\tilde{U}$ is given by

$$\tilde{U} = \tilde{Z}_1 - \frac{\tilde{\lambda}_3}{2\tilde{\lambda}_2^{3/2}\sqrt{n}}\tilde{Z}_1^2 + \frac{1}{2n}\tilde{Z}_1 + \frac{1}{2n}\left(\frac{\tilde{\lambda}_3^2}{\tilde{\lambda}_2^3} - \frac{\tilde{\lambda}_4}{3\tilde{\lambda}_2^2}\right)\tilde{Z}_1^3 + O_p\left(\frac{1}{n\sqrt{n}}\right). \quad (2.31)$$

Since

$$\tilde{\lambda}_2 = \lambda_2(\theta, X_{(1)}) = \lambda_2(\theta, \gamma) + \frac{1}{n}\left(\frac{\partial \lambda_2}{\partial \gamma}\right)T_{(1)} + O_p\left(\frac{1}{n^2}\right),$$

we obtain

$$\tilde{U} = \sqrt{\lambda_2 n}(\hat{\theta}_{MCL} - \theta)\left\{1 + \frac{1}{2n\lambda_2}\left(\frac{\partial \lambda_2}{\partial \gamma}\right)T_{(1)} + O_p\left(\frac{1}{n^2}\right)\right\}, \quad (2.32)$$

where $T_{(1)} = n(X_{(1)} - \gamma)$ and $\lambda_2 = \lambda_2(\theta, \gamma)$. Then, it follows from (2.31) and (2.32) that

$$\begin{aligned}\tilde{U}_0 &= \sqrt{\lambda_2 n}(\hat{\theta}_{MCL} - \theta)\\ &= \tilde{Z}_1 - \frac{\tilde{\lambda}_3}{2\tilde{\lambda}_2^{3/2}\sqrt{n}}\tilde{Z}_1^2 + \frac{1}{2n}\left\{1 - \frac{1}{\lambda_2}\left(\frac{\partial \lambda_2}{\partial \gamma}\right)T\right\}\tilde{Z}_1\\ &\quad + \frac{1}{2n}\left(\frac{\tilde{\lambda}_3^2}{\tilde{\lambda}_2^3} - \frac{\tilde{\lambda}_4}{3\tilde{\lambda}_2^2}\right)\tilde{Z}_1^3 + O_p\left(\frac{1}{n\sqrt{n}}\right). \end{aligned} \quad (2.33)$$

For given $X_{(1)} = x_{(1)}$, i.e., $T_{(1)} = t := n(x_{(1)} - \gamma)$, the conditional expectation of $\tilde{Z}_1$ and $\tilde{Z}_1^2$ are

$$E_{\theta,\gamma}(\tilde{Z}_1|t) = \frac{1}{\sqrt{\tilde{\lambda}_2(n-1)}} \sum_{i=2}^{n} \{E_{\theta,\gamma}[u(Y_i)|t] - \lambda_1(\theta, x_{(1)})\} = 0,$$

$$\begin{aligned} E_{\theta,\gamma}(\tilde{Z}_1^2|t) = &\frac{1}{\tilde{\lambda}_2(n-1)}\Bigg[\sum_{i=2}^{n} E_{\theta,\gamma}[\{u(Y_i) - \lambda_1(\theta, x_{(1)})\}^2|t] \\ &+ \sum_{\substack{i \neq j \\ 2 \le i,j \le n}}\sum E_{\theta,\gamma}\left[\{u(Y_i) - \lambda_1(\theta, x_{(1)})\}\{u(Y_j) - \lambda_1(\theta, x_{(1)})\} \,|\, t\right]\Bigg] \\ = &\, 1, \end{aligned} \tag{2.34}$$

hence, the conditional variance of $\tilde{Z}_1$ is equal to 1, i.e., $V_{\theta,\gamma}(\tilde{Z}_1|t) = 1$. In a similar way to the above, we have

$$E_{\theta,\gamma}(\tilde{Z}_1^3|t) = \frac{\tilde{\lambda}_3}{\tilde{\lambda}_2^{3/2}\sqrt{n-1}}, \quad E_{\theta,\gamma}(\tilde{Z}_1^4|t) = 3 + \frac{\tilde{\lambda}_4}{\tilde{\lambda}_2^2(n-1)}. \tag{2.35}$$

Then, it follows from (2.33)–(2.35) that

$$E_{\theta,\gamma}(\tilde{U}_0|T_{(1)}) = -\frac{\tilde{\lambda}_3}{2\tilde{\lambda}_2^{3/2}\sqrt{n}} + O_p\left(\frac{1}{n\sqrt{n}}\right), \tag{2.36}$$

$$\begin{aligned} E_{\theta,\gamma}(\tilde{U}_0^2|T_{(1)}) = &\,1 + \frac{1}{n} + \frac{1}{n}\left(\frac{11\tilde{\lambda}_3^2}{4\tilde{\lambda}_2^3} - \frac{\tilde{\lambda}_4}{\tilde{\lambda}_2^2}\right) - \frac{1}{\lambda_2 n}\left(\frac{\partial \lambda_2}{\partial \gamma}\right) T_{(1)} \\ &+ O_p\left(\frac{1}{n\sqrt{n}}\right), \end{aligned} \tag{2.37}$$

where $\tilde{\lambda}_i = \lambda_i(\theta, X_{(1)})$ $(i = 2, 3, 4)$. Since, for $i = 2, 3, 4$

$$\tilde{\lambda}_i = \lambda_i(\theta, X_{(1)}) = \lambda_i(\theta, \gamma) + O_p\left(\frac{1}{n}\right) = \lambda_i + O_p\left(\frac{1}{n}\right), \tag{2.38}$$

it follows from (2.36) that

$$E_{\theta,\gamma}(\tilde{U}_0) = -\frac{\lambda_3}{2\lambda_2^{3/2}\sqrt{n}} + O\left(\frac{1}{n\sqrt{n}}\right). \tag{2.39}$$

It is noted from (2.12), (2.26), and (2.39) that

$$E_{\theta,\gamma}(U_\gamma) = E_{\theta,\gamma}(\hat{U}^*) = E_{\theta,\gamma}(\tilde{U}_0) = -\frac{\lambda_3}{2\lambda_2^{3/2}\sqrt{n}} + O\left(\frac{1}{n\sqrt{n}}\right).$$

In a similar way to the above, we obtain from (2.16), (2.37), and (2.38)

$$E_{\theta,\gamma}(\tilde{U}_0^2) = 1 + \frac{1}{n} + \frac{11\lambda_3^2}{4\lambda_2^3 n} - \frac{\lambda_4}{\lambda_2^2 n} - \frac{1}{k\lambda_2 n}\left(\frac{\partial \lambda_2}{\partial \gamma}\right) + O\left(\frac{1}{n\sqrt{n}}\right). \tag{2.40}$$

Since, by (2.28) and (2.29)

$$\frac{1}{k}\left(\frac{\partial \lambda_2}{\partial \gamma}\right) = \frac{1}{k}\left\{\frac{\partial k}{\partial \theta}(\lambda_1 - u(\gamma)) + k\left(\frac{\partial \lambda_1}{\partial \theta}\right)\right\} = -(\lambda_1 - u(\gamma))^2 + \lambda_2,$$

it follows from (2.40) that

$$E_{\theta,\gamma}(\tilde{U}_0^2) = 1 + \frac{11\lambda_3^2}{4\lambda_2^3 n} - \frac{\lambda_4}{\lambda_2^2 n} + \frac{1}{\lambda_2 n}\{\lambda_1 - u(\gamma)\}^2 + O\left(\frac{1}{n\sqrt{n}}\right),$$

hence, by (2.39)

$$V_{\theta,\gamma}(\tilde{U}_0) = 1 + \frac{1}{n}\left(\frac{5\lambda_3^2}{2\lambda_2^3} - \frac{\lambda_4}{\lambda_2^2}\right) + \frac{1}{\lambda_2 n}\{\lambda_1 - u(\gamma)\}^2 + O\left(\frac{1}{n\sqrt{n}}\right). \tag{2.41}$$

From (2.33), (2.39) and (2.41), we have the conclusion of Theorem 2.5.1.

2.10 Appendix A2

The proof of Lemma 2.9.1 Since the second-order asymptotic cumulative distribution function of $T_{(1)}$ is given by

$$\begin{aligned} F_{T_{(1)}}(t) &= P_{\theta,\gamma}\left\{T_{(1)} \le t\right\} = P_{\theta,\gamma}\left\{n(X_{(1)} - \gamma) \le t\right\} = 1 - \left\{\frac{b(\theta, \gamma + (t/n))}{b(\theta,\gamma)}\right\}^n \\ &= 1 - e^{-k(\theta,\gamma)t}\left[1 - \frac{t^2}{2n}\left\{\frac{\partial k(\theta,\gamma)}{\partial \gamma}\right\} + O\left(\frac{1}{n^2}\right)\right] \end{aligned}$$

for $t > 0$, we obtain (2.15). From (2.15), we also get (2.16) by a straightforward calculation.

The proof of Lemma 2.9.2 As is seen from the beginning of Sect. 2.5, $Y_2, \ldots, Y_n$ are i.i.d. random variables according to a distribution with density

$$g(y; \theta, x_{(1)}) = \frac{a(y)e^{\theta u(y)}}{b(\theta, x_{(1)})} \quad \text{for } x_{(1)} \le y < d \tag{2.42}$$

with respect to the Lebesgue measure. Then, the conditional expectation of Z_1 given $T_{(1)}$ is obtained by

$$E_{\theta,\gamma}(Z_1|T_{(1)}) = \frac{1}{\sqrt{\lambda_2 n}}\left\{u(X_{(1)}) + \sum_{i=2}^{n} E_{\theta,\gamma}[u(Y_i)|T_{(1)}] - n\lambda_1\right\}, \tag{2.43}$$

where $\lambda_i = \lambda_i(\theta, \gamma)$ $(i = 1, 2)$. Since, for each $i = 2, \ldots, n$, by (2.42)

$$E_{\theta,\gamma}[u(Y_i)|T_{(1)}] = \frac{\partial}{\partial\theta}\log b(\theta, X_{(1)}) = \lambda_1(\theta, X_{(1)}) =: \hat{\lambda}_1 \quad \text{(say)},$$

it follows from (2.43) that

$$E_{\theta,\gamma}(Z_1|T_{(1)}) = \frac{1}{\sqrt{\lambda_2 n}}\left\{u(X_{(1)}) + (n-1)\hat{\lambda}_1\right\} - \frac{\lambda_1\sqrt{n}}{\sqrt{\lambda_2}},$$

hence, from (2.16) and (2.43)

$$\begin{aligned} E_{\theta,\gamma}(Z_1 T_{(1)}) = &\frac{1}{\sqrt{\lambda_2 n}}\left\{E_{\theta,\gamma}[u(X_{(1)})T_{(1)}] + (n-1)E_{\theta,\gamma}(\hat{\lambda}_1 T_{(1)})\right\} \\ &- \sqrt{\frac{n}{\lambda_2}}\lambda_1\left\{\frac{1}{k} + \frac{A(\theta,\gamma)}{n} + O\left(\frac{1}{n^2}\right)\right\}, \end{aligned} \tag{2.44}$$

where $k = k(\theta, \gamma)$. Since, by the Taylor expansion

$$\begin{aligned} u(X_{(1)}) &= u(\gamma) + \frac{u'(\gamma)}{n}T_{(1)} + \frac{u''(\gamma)}{2n^2}T_{(1)}^2 + O_p\left(\frac{1}{n^3}\right), \\ \hat{\lambda}_1 &= \lambda_1(\theta, X_{(1)}) = \lambda_1(\theta,\gamma) + \frac{1}{n}\left\{\frac{\partial}{\partial\gamma}\lambda_1(\theta,\gamma)\right\}T_{(1)} \\ &\quad + \frac{1}{2n^2}\left\{\frac{\partial^2}{\partial\gamma^2}\lambda_1(\theta,\gamma)\right\}T_{(1)}^2 + O_p\left(\frac{1}{n^3}\right), \end{aligned}$$

it follows from (2.16) that

$$E_{\theta,\gamma}[u(X_{(1)})T_{(1)}] = \frac{u(\gamma)}{k} + \frac{1}{n}\left\{Au(\gamma) + \frac{2u'(\gamma)}{k^2}\right\} + O\left(\frac{1}{n^2}\right), \tag{2.45}$$

$$E_{\theta,\gamma}(\hat{\lambda}_1 T_{(1)}) = \frac{\lambda_1}{k} + \frac{1}{n}\left\{\lambda_1 A + \frac{2}{k^2}\left(\frac{\partial\lambda_1}{\partial\gamma}\right)\right\} + O\left(\frac{1}{n^2}\right), \tag{2.46}$$

where $k = k(\theta, \gamma)$, $A = A(\theta, \gamma)$, and $\lambda_1 = \lambda_1(\theta, \gamma)$. From (2.44)–(2.46), we obtain (2.17).

The proof of Lemma 2.9.3 First, we have

$$E_{\theta,\gamma}(Z_1^2|T_{(1)}) = \frac{1}{\lambda_2 n}\left\{u(X_{(1)}) - \lambda_1\right\}^2 + \frac{2}{\lambda_2 n}\left\{u(X_{(1)}) - \lambda_1\right\}\sum_{i=2}^{n} E_{\theta,\gamma}\left[u(Y_i) - \lambda_1|T_{(1)}\right] + \frac{1}{\lambda_2 n}\sum_{i=2}^{n} E_{\theta,\gamma}\left[\{u(Y_i) - \lambda_1\}^2|T_{(1)}\right] + \frac{1}{\lambda_2 n}\sum\sum_{\substack{i\neq j \\ 2\le i,j\le n}} E_{\theta,\gamma}\left[\{u(Y_i) - \lambda_1\}\{u(Y_j) - \lambda_1\} \mid T_{(1)}\right]. \tag{2.47}$$

For $2 \le i \le n$, we have

$$E_{\theta,\gamma}[u(Y_i) - \lambda_1|T_{(1)}] = \left(\frac{\partial\lambda_1}{\partial\gamma}\right)\frac{T_{(1)}}{n} + O_p\left(\frac{1}{n^2}\right) = O_p\left(\frac{1}{n}\right), \tag{2.48}$$

and for $i \neq j$ and $2 \le i, j \le n$

$$E_{\theta,\gamma}\left[\{u(Y_i) - \lambda_1\}\{u(Y_j) - \lambda_1\}|T_{(1)}\right] = E_{\theta,\gamma}\left[u(Y_i) - \lambda_1 \mid T_{(1)}\right]E_{\theta,\gamma}\left[u(Y_j) - \lambda_1|T_{(1)}\right] = \left(\frac{\partial\lambda_1}{\partial\gamma}\right)^2\frac{T_{(1)}^2}{n^2} + O_p\left(\frac{1}{n^3}\right) = O_p\left(\frac{1}{n^2}\right). \tag{2.49}$$

Since, for $i = 2, \ldots, n$

$$E_{\theta,\gamma}[u^2(Y_i)|T_{(1)}] = \hat{\lambda}_1^2 + \hat{\lambda}_2,$$

where $\hat{\lambda}_i = \lambda_i(\theta, X_{(1)})$ $(i = 1, 2)$, we have for $i = 2, \ldots, n$

$$E_{\theta,\gamma}[\{u(Y_i) - \lambda_1\}^2|T_{(1)}] = \lambda_2 + \frac{1}{n}\left(\frac{\partial\lambda_2}{\partial\gamma}\right)T_{(1)} + O_p\left(\frac{1}{n^2}\right) = \lambda_2 + O_p\left(\frac{1}{n}\right). \tag{2.50}$$

From (2.47)–(2.50), we obtain

$$E_{\theta,\gamma}(Z_1^2|T_{(1)}) = 1 + O_p\left(\frac{1}{n}\right),$$

hence, by (2.16)

$$E_{\theta,\gamma}(Z_1^2 T_{(1)}) = E_{\theta,\gamma}[T_{(1)} E_{\theta,\gamma}(Z_1^2 | T_{(1)})] = E_{\theta,\gamma}(T_{(1)}) + O\left(\frac{1}{n}\right) = \frac{1}{k} + O\left(\frac{1}{n}\right).$$

Thus, we get (2.18).

References

Akahira, M. (1986). *The structure of asymptotic deficiency of estimators*. Queen's Papers in Pure and Applied Mathematics (Vol. 75). Kingston, Canada: Queen's University Press.

Akahira, M., & Ohyauchi, N. (2012). The asymptotic expansion of the maximum likelihood estimator for a truncated exponential family of distributions. (In Japanese). *Kôkyûroku*: RIMS (Research Institute for Mathematical Sciences), Kyoto University, (No. 1804, pp. 188–192).

Akahira, M. (2016). Second order asymptotic comparison of the MLE and MCLE of a natural parameter for a truncated exponential family of distributions. *Annals of the Institute of Statistical Mathematics*, *68*, 469–490.

Akahira, M., & Takeuchi, K. (1982). On asymptotic deficiency of estimators in pooled samples in the presence of nuisance parameters. *Statistics & Decisions*, *1*, 17–38.

Andersen, E. B. (1970). Asymptotic properties of conditional maximum likelihood estimators. *Journal of the Royal Statistical Society Series B*, *32*, 283–301.

Arnold, B. C. (2015). *Pareto distributions* (2nd ed.). Boca Raton: CRC Press.

Bar-Lev, S. K. (1984). Large sample properties of the MLE and MCLE for the natural parameter of a truncated exponential family. *Annals of the Institute of Statistical Mathematics*, *36*, Part A, 217–222.

Bar-Lev, S. K., & Reiser, B. (1983). A note on maximum conditional likelihood estimation. *Sankhyā Series A*, *45*, 300–302.

Barndorff-Nielsen, O. E. (1978). *Information and exponential families in statistical theory*. New York: Wiley.

Basu, D. (1977). On the elimination of nuisance parameters. *Journal of the American Statistical Association*, *72*, 355–366.

Cox, D. R., & Reid, N. (1987). Parameter orthogonality and approximate conditional inference (with discussion). *Journal of the Royal Statistical Society Series B*, *49*, 1–39.

Ferguson, H. (1992). Asymptotic properties of a conditional maximum-likelihood estimator. *Canadian Journal of Statistics*, *20*, 63–75.

Huque, F., & Katti, S. K. (1976). A note on maximum conditional estimators. *Sankhyā Series B*, *38*, 1–13.

Liang, K.-Y. (1984). The asymptotic efficiency of conditional likelihood methods. *Biometrika*, *71*, 305–313.

Lwin, T. (1975). Exponential family distribution with a truncation parameter. *Biometrika*, *62*, 218–220.

Quesenberry, C. P. (1975). Transforming samples from truncation parameter distributions to uniformity. *Communications in Statistics*, *4*, 1149–1155.

Vancak, V., Goldberg, Y., Bar-Lev, S. K., & Boukai, B. (2015). Continuous statistical models: With or without truncation parameters? *Mathematical Methods of Statistics*, *24*, 55–73.

Voinov, V. G., & Nikulin, M. S. (1993). *Unbiased estimators and their applications* (Vol. 1) Univariate Case. Dordrecht: Kluwer Academic Publishers.

Chapter 3
Maximum Likelihood Estimation of a Natural Parameter for a Two-Sided TEF

The corresponding results on a comparison of the estimators of a natural parameter θ to the case of a oTEF in the previous chapter are obtained in the case of a two-sided truncated exponential family (tTEF) of distributions with a natural parameter θ and two truncation parameters γ and ν as nuisance ones.

3.1 Introduction

In the previous chapter, for a oTEF of distributions with a natural parameter θ and a truncation parameter γ as a nuisance parameter, the second-order asymptotic losses of $\hat{\theta}_{ML^*}$ and $\hat{\theta}_{MCL}$ relative to $\hat{\theta}^{\gamma}_{ML}$ which correspond to the asymptotic deficiencies are obtained from their second-order asymptotic variances which are calculated from their stochastic expansions. It is also shown that a bias-adjusted MLE $\hat{\theta}_{ML^*}$ and $\hat{\theta}_{MCL}$ of θ for unknown γ are second-order asymptotically equivalent and second-order asymptotically worse than $\hat{\theta}^{\gamma}_{ML}$ of θ for known γ. On the other hand, for an upper-truncated Pareto distribution with an index parameter α to be estimated and two truncation parameters γ and ν as nuisance ones, the MLE $\tilde{\alpha}$ of α for known γ and ν and the MLE $\hat{\alpha}$ of α for unknown γ and ν are shown to have the asymptotic normality by Aban et al. (2006). The distribution does not belong to oTEF but to a tTEF of distributions.

In this chapter, following mostly the paper by Akahira et al. (2016), the corresponding results on the second-order asymptotic comparison of the estimators of θ to the case of oTEF in Chap. 2 are obtained in the case of a tTEF of distributions with a natural parameter θ and two truncation parameters γ and ν as nuisance ones. The upper-truncated Pareto case is treated in Example 3.7.3.

M. Akahira, *Statistical Estimation for Truncated Exponential Families*,
JSS Research Series in Statistics, DOI 10.1007/978-981-10-5296-5_3

3.2 Preliminaries

Suppose that $X_1, X_2, \ldots, X_n, \ldots$ is a sequence of i.i.d. random variables according to $P_{\theta,\gamma,\nu}$ in a tTEF $\mathscr{P}_t$ with the density (1.9). Then, we consider the estimation problem on the natural parameter θ in the presence of nuisance parameters γ and ν. Denote a random vector $(X_1, \ldots, X_n)$ by $\boldsymbol{X}$, and let $X_{(1)} \leq \cdots \leq X_{(n)}$ be the corresponding order statistics of a random vector $\boldsymbol{X}$. Then, the density (1.9) is considered to belong to a regular exponential family of distributions with a natural parameter θ for any fixed γ and ν; hence, $\log b(\theta, \gamma, \nu)$ is strictly convex and infinitely differentiable in $\theta \in \Theta$ and

$$\lambda_j(\theta, \gamma, \nu) := \frac{\partial^j}{\partial \theta^j} \log b(\theta, \gamma, \nu) \tag{3.1}$$

is the jth cumulant corresponding to (1.9) for $j = 1, 2, \ldots$.

In Sects. 3.3 to 3.5, the stochastic expansions of the MLE $\hat{\theta}_{ML}^{\gamma,\nu}$ of θ for known γ and ν, the MLE $\hat{\theta}_{ML}$ and the MCLE $\hat{\theta}_{MCL}$ of θ for unknown γ and ν are derived, from which the second-order asymptotic means and variances are obtained. In Sect. 3.6, the second-order asymptotic losses of $\hat{\theta}_{ML^*}$ and $\hat{\theta}_{MCL}$ relative to $\hat{\theta}_{ML}^{\gamma,\nu}$ is obtained from their second-order asymptotic variances, and a bias-adjusted MLE $\hat{\theta}_{ML^*}$ and $\hat{\theta}_{MCL}$ of θ for unknown γ and ν are also shown to be second-order asymptotically equivalent and second-order asymptotically worse than the MLE $\hat{\theta}_{ML}^{\gamma,\nu}$ for known γ and ν. In Sect. 3.7, examples for a two-sided truncated exponential, a two-sided truncated normal, an upper-truncated Pareto, a two-sided truncated beta, a two-sided truncated Erlang type, and a two-sided truncated lognormal distributions are given. In Appendices B1 and B2, the proofs of theorems are given.

3.3 MLE $\hat{\theta}_{ML}^{\gamma,\nu}$ of θ When Truncation Parameters γ and ν are Known

For given $\boldsymbol{x} = (x_1, \ldots, x_n)$ satisfying $c < \gamma \leq x_{(1)} := \min_{1 \leq i \leq n} x_i$ and $x_{(n)} := \max_{1 \leq i \leq n} x_i \leq \nu < d$, the likelihood function of θ is given by

$$L^{\gamma,\nu}(\theta; \boldsymbol{x}) := \frac{1}{b^n(\theta, \gamma, \nu)} \left\{ \prod_{i=1}^n a(x_i) \right\} \exp\left\{ \theta \sum_{i=1}^n u(x_i) \right\}.$$

Then, the likelihood equation is

$$\frac{1}{n} \sum_{i=1}^n u(x_i) - \lambda_1(\theta, \gamma, \nu) = 0. \tag{3.2}$$

Since there exists a unique solution of the Eq. (3.2) with respect to θ, we denote it by $\hat{\theta}_{ML}^{\gamma,\nu}$ which is the MLE of θ. Let $\lambda_i = \lambda_i(\theta, \gamma, \nu)$ $(i = 2, 3, 4)$ and put

$$Z_1 := \frac{1}{\sqrt{\lambda_2 n}} \sum_{i=1}^{n} \{u(X_i) - \lambda_1\}, \quad U_{\gamma,\nu} := \sqrt{\lambda_2 n}(\hat{\theta}_{ML}^{\gamma,\nu} - \theta).$$

Then, we have the following.

Theorem 3.3.1 *For a tTEF $\mathscr{P}_t$ of distributions with densities of the form (1.9) with a natural parameter θ and truncation parameters γ and ν, let $\hat{\theta}_{ML}^{\gamma,\nu}$ be the MLE of θ when γ and ν are known. Then, the stochastic expansion of $U_{\gamma,\nu}$ is given by*

$$U_{\gamma,\nu} = Z_1 - \frac{\lambda_3}{2\lambda_2^{3/2}\sqrt{n}} Z_1^2 + \frac{1}{2n}\left(\frac{\lambda_3^2}{\lambda_2^3} - \frac{\lambda_4}{3\lambda_2^2}\right) Z_1^3 + O_p\left(\frac{1}{n\sqrt{n}}\right),$$

and the second-order asymptotic mean and variance are given by

$$E_\theta(U_{\gamma,\nu}) = -\frac{\lambda_3}{2\lambda_2^{3/2}\sqrt{n}} + O\left(\frac{1}{n\sqrt{n}}\right),$$

$$V_\theta(U_{\gamma,\nu}) = 1 + \frac{1}{n}\left(\frac{5\lambda_3^2}{2\lambda_2^3} - \frac{\lambda_4}{\lambda_2^2}\right) + O\left(\frac{1}{n\sqrt{n}}\right),$$

respectively.

The proof of Theorem 3.3.1 is omitted, since it is similar to that of Theorem 2.3.1. Since $U_{\gamma,\nu} = Z_1 + o_p(1)$, it is seen that $U_{\gamma,\nu}$ is asymptotically normal with mean 0 and variance 1.

3.4 Bias-Adjusted MLE $\hat{\theta}_{ML^*}$ of θ When γ and ν are Unknown

For given $\boldsymbol{x} = (x_1, \dots, x_n)$ satisfying $c < \gamma \le x_{(1)}$ and $x_{(n)} \le \nu < d$, the likelihood function of θ, γ, and ν is given by

$$L(\theta, \gamma, \nu; \boldsymbol{x}) = \frac{1}{b^n(\theta, \gamma, \nu)} \left\{\prod_{i=1}^{n} a(x_i)\right\} \exp\left\{\theta \sum_{i=1}^{n} u(x_i)\right\}. \tag{3.3}$$

Let $\hat{\theta}_{ML}$, $\hat{\gamma}_{ML}$, and $\hat{\nu}_{ML}$ be the MLEs of θ, γ and ν, respectively. Then, it follows from (3.3) that $\hat{\gamma}_{ML} = X_{(1)}$ and $\hat{\nu}_{ML} = X_{(n)}$ and $L(\hat{\theta}_{ML}, X_{(1)}, X_{(n)}; \boldsymbol{X}) = \sup_{\theta \in \Theta} L(\theta, X_{(1)}, X_{(n)}; \boldsymbol{X})$, hence $\hat{\theta}_{ML}$ satisfies the likelihood equation

$$0 = \frac{1}{n}\sum_{i=1}^{n} u(X_i) - \lambda_1(\hat{\theta}_{ML}, X_{(1)}, X_{(n)}) \tag{3.4}$$

where $\boldsymbol{X} = (X_1, \ldots, X_n)$. Put $\hat{U} := \sqrt{\lambda_2 n}(\hat{\theta}_{ML} - \theta)$, $T_{(1)} := n(X_{(1)} - \gamma)$, $T_{(n)} := n(X_{(n)} - \nu)$. Then, we have the following.

Theorem 3.4.1 *For a tTEF $\mathscr{P}_t$ of distributions with densities of the form (1.9) with a natural parameter θ and truncation parameters γ and ν, let $\hat{\theta}_{ML}$ be the MLE of θ when γ and ν are unknown, and $\hat{\theta}_{ML^*}$ be a bias-adjusted MLE such that $\hat{\theta}_{ML}$ has the same asymptotic bias as that $\hat{\theta}_{ML}^{\gamma,\nu}$, i.e.,*

$$\begin{aligned}\hat{\theta}_{ML^*} = \hat{\theta}_{ML} + \frac{1}{\lambda_2(\hat{\theta}_{ML}, X_{(1)}, X_{(n)})}\Bigg\{&\frac{1}{k(\hat{\theta}_{ML}, X_{(1)}, X_{(n)})}\frac{\partial\lambda_1}{\partial\gamma}(\hat{\theta}_{ML}, X_{(1)}, X_{(n)})\\ &- \frac{1}{\tilde{k}(\hat{\theta}_{ML}, X_{(1)}, X_{(n)})}\frac{\partial\lambda_1}{\partial\nu}(\hat{\theta}_{ML}, X_{(1)}, X_{(n)})\Bigg\}\end{aligned} \tag{3.5}$$

where $k(\theta, \gamma, \nu) := a(\gamma)e^{\theta u(\gamma)}/b(\theta, \gamma, \nu)$ and $\tilde{k}(\theta, \gamma, \nu) := a(\nu)e^{\theta u(\nu)}/b(\theta, \gamma, \nu)$. Then, the stochastic expansion of $\hat{U}^* := \sqrt{\lambda_2 n}(\hat{\theta}_{ML^*} - \theta)$ is given by

$$\begin{aligned}\hat{U}^* = \hat{U} &+ \frac{1}{\sqrt{\lambda_2 n}}\left\{\frac{1}{k}\left(\frac{\partial\lambda_1}{\partial\gamma}\right) - \frac{1}{\tilde{k}}\left(\frac{\partial\lambda_1}{\partial\nu}\right)\right\}\\ &- \frac{1}{\lambda_2 n}\left\{\frac{\delta_1}{k} - \frac{\delta_2}{\tilde{k}} + \frac{1}{k^2}\left(\frac{\partial k}{\partial\theta}\right)\left(\frac{\partial\lambda_1}{\partial\gamma}\right) - \frac{1}{\tilde{k}^2}\left(\frac{\partial\tilde{k}}{\partial\theta}\right)\left(\frac{\partial\lambda_1}{\partial\nu}\right)\right\}Z_1 + O_p\left(\frac{1}{n\sqrt{n}}\right),\end{aligned} \tag{3.6}$$

with

$$\begin{aligned}\hat{U} = Z_1 &- \frac{\lambda_3}{2\lambda_2^{3/2}\sqrt{n}}Z_1^2 - \frac{1}{\sqrt{\lambda_2 n}}\left\{\left(\frac{\partial\lambda_1}{\partial\gamma}\right)T_{(1)} + \left(\frac{\partial\lambda_1}{\partial\nu}\right)T_{(n)}\right\}\\ &+ \frac{1}{\lambda_2 n}Z_1\{\delta_1 T_{(1)} + \delta_2 T_{(n)}\} + \frac{1}{2n}\left(\frac{\lambda_3^2}{\lambda_2^3} - \frac{\lambda_4}{3\lambda_2^2}\right)Z_1^3 + O_p\left(\frac{1}{n\sqrt{n}}\right)\end{aligned} \tag{3.7}$$

where $k = k(\theta, \gamma, \nu)$, $\tilde{k} = \tilde{k}(\theta, \gamma, \nu)$, and

$$\delta_1 := \frac{\lambda_3}{\lambda_2}\left(\frac{\partial\lambda_1}{\partial\gamma}\right) - \frac{\partial\lambda_2}{\partial\gamma}, \quad \delta_2 := \frac{\lambda_3}{\lambda_2}\left(\frac{\partial\lambda_1}{\partial\nu}\right) - \frac{\partial\lambda_2}{\partial\nu}, \tag{3.8}$$

and the second-order asymptotic mean and variance of $\hat{U}^*$ are given by

$$E_{\theta,\gamma,\nu}(\hat{U}^*) = -\frac{\lambda_3}{2\lambda_2^{3/2}\sqrt{n}} + o\left(\frac{1}{n\sqrt{n}}\right), \tag{3.9}$$

$$V_{\theta,\gamma,\nu}(\hat{U}^*) = 1 + \frac{1}{n}\left(\frac{5\lambda_3^2}{2\lambda_2^3} - \frac{\lambda_4}{\lambda_2^2}\right) + \frac{1}{\lambda_2 n}\left[\{u(\gamma)-\lambda_1\}^2 + \{u(\nu)-\lambda_1\}^2\right] + o\left(\frac{1}{n\sqrt{n}}\right), \tag{3.10}$$

respectively.

Since $\hat{U} = \hat{U}^* = Z_1 + o_p(1)$, it seen that $\hat{U}$ and $\hat{U}^*$ are asymptotically normal with mean 0 and variance 1 in the first order. It is also noted from Theorems 3.3.1 and 3.4.1 that there is a difference between $V_\theta(U_{\gamma,\nu})$ and $V_{\theta,\gamma,\nu}(\hat{U}^*)$ in the second order, i.e., the order n^{-1}. Further, it is remarked that the asymptotic distribution of $T_{(1)}$ and $T_{(n)}$ are exponential and given up to the second order (see Lemmas 3.9.1 and 3.9.2 in Appendix B1).

3.5 MCLE $\hat{\theta}_{MCL}$ of θ When γ and ν are Unknown

Let $Y_2, \ldots, Y_{n-1}$ be a random permutation of the $(n-2)!$ permutations of $X_{(2)}, \ldots, X_{(n-1)}$ such that conditionally on $X_{(1)} = x_{(1)}$ and $X_{(n)} = x_{(n)}$, $Y_2, \ldots, Y_{n-1}$ are i.i.d. random variables according to a distribution with density

$$f(y;\theta, x_{(1)}, x_{(n)}) = \frac{a(y)e^{\theta u(y)}}{b(\theta, x_{(1)}, x_{(n)})} \quad \text{for} \quad c < \gamma \le x_{(1)} < y < x_{(n)} \le \nu < d \tag{3.11}$$

with respect to the Lebesgue measure. For given $X_{(1)} = x_{(1)}$ and $X_{(n)} = x_{(n)}$, the conditional likelihood function of θ for $\mathbf{y} = (y_2, \ldots, y_{n-1})$ satisfying $c < \gamma \le x_{(1)} \le y_i \le x_{(n)} \le \nu < d$ $(i = 2, \ldots, n-1)$ is

$$L(\theta; \mathbf{y}|x_{(1)}, x_{(n)}) = \frac{1}{b^{n-2}(\theta, x_{(1)}, x_{(n)})}\left\{\prod_{i=2}^{n-1} a(y_i)\right\} \exp\left\{\theta \sum_{i=2}^{n-1} u(y_i)\right\}.$$

Then, the likelihood equation is

$$\frac{1}{n-2}\sum_{i=2}^{n-1} u(y_i) - \lambda_1(\theta, x_{(1)}, x_{(n)}) = 0. \tag{3.12}$$

Since there exists a unique solution of the Eq. (3.12) with respect to θ, we denote it by $\hat{\theta}_{MCL}$, i.e., the value of θ for which $L(\theta; \mathbf{y}|x_{(1)}, x_{(n)})$ attains supremum. Let $\tilde{\lambda}_i = \tilde{\lambda}_i(\theta, x_{(1)}, x_{(n)})$ $(i = 1, 2, 3, 4)$ and put

$$\tilde{Z}_1 := \frac{1}{\sqrt{\tilde{\lambda}_2(n-2)}} \sum_{i=2}^{n-1} \{u(Y_i) - \tilde{\lambda}_1\}, \quad \tilde{U}_0 := \sqrt{\tilde{\lambda}_2 n}(\hat{\theta}_{MCL} - \theta).$$

Then, we have the following.

Theorem 3.5.1 *For a tTEF $\mathscr{P}_t$ of distributions with densities of the form (1.9) with a natural parameter θ and truncation parameters γ and ν, let $\hat{\theta}_{MCL}$ be the MCLE of θ when γ and ν are unknown. Then, the stochastic expansion of $\tilde{U}_0$ is given by*

$$\tilde{U}_0 = \tilde{Z}_1 - \frac{\tilde{\lambda}_3}{2\tilde{\lambda}_2^{3/2}\sqrt{n}} \tilde{Z}_1^2 + \frac{1}{n}\left\{1 - \frac{1}{2\lambda_2}\left(\frac{\partial \lambda_2}{\partial \gamma}\right) T_{(1)} - \frac{1}{2\lambda_2}\left(\frac{\partial \lambda_2}{\partial \nu}\right) T_{(n)}\right\} \tilde{Z}_1$$
$$+ \frac{1}{2n}\left(\frac{\tilde{\lambda}_3^2}{\tilde{\lambda}_2^3} - \frac{\tilde{\lambda}_4}{3\tilde{\lambda}_2^2}\right) \tilde{Z}_1^3 + O_p\left(\frac{1}{n\sqrt{n}}\right) \tag{3.13}$$

and the second-order asymptotic mean and variance are given by

$$E_{\theta,\gamma,\nu}(\tilde{U}_0) = -\frac{\lambda_3}{2\lambda_2^{3/2}\sqrt{n}} + O\left(\frac{1}{n\sqrt{n}}\right),$$
$$V_{\theta,\gamma,\nu}(\tilde{U}_0) = 1 + \frac{1}{n}\left(\frac{5\lambda_3^2}{2\lambda_2^3} - \frac{\lambda_4}{\lambda_2^2}\right) + \frac{1}{\lambda_2 n}\left[\{u(\gamma) - \lambda_1\}^2 + \{u(\nu) - \lambda_1\}^2\right]$$
$$+ O\left(\frac{1}{n\sqrt{n}}\right),$$

respectively.

Theorems 3.4.1 and 3.5.1 also lead to a similar explanation to Remarks 2.5.1 and 2.5.2 in Chap. 2.

3.6 Second-Order Asymptotic Comparison Among $\hat{\theta}_{ML}^{\gamma,\nu}$, $\hat{\theta}_{ML^*}$, and $\hat{\theta}_{MCL}$

From the results in the previous sections, we have the following.

Theorem 3.6.1 *For a tTEF $\mathscr{P}_t$ of distributions with densities of the form (1.9) with a natural parameter θ and truncation parameters γ and ν, let $\hat{\theta}_{ML}^{\gamma,\nu}$, $\hat{\theta}_{ML^*}$, and $\hat{\theta}_{MCL}$ be the MLE of θ when γ and ν are known, the bias-adjusted MLE and the MCLE of θ when γ and ν are unknown, respectively. Then, the bias-adjusted MLE $\hat{\theta}_{ML^*}$ and* *the MCLE $\hat{\theta}_{MCL}$ are second-order asymptotically equivalent in the sense that*

$$d_n(\hat{\theta}_{ML^*}, \hat{\theta}_{MCL}) := n\left\{V_{\theta,\gamma,\nu}(\hat{U}^*) - V_{\theta,\gamma,\nu}(\tilde{U}_0)\right\} = o(1)$$

as $n \to \infty$, and they are second-order asymptotically worse than $\hat{\theta}_{ML}^{\gamma,\nu}$ with the following second-order asymptotic losses of $\hat{\theta}_{ML^*}$and $\hat{\theta}_{MCL}$ relative to $\hat{\theta}_{ML}^{\gamma,\nu}$:

$$\begin{aligned} d_n(\hat{\theta}_{ML^*}, \hat{\theta}_{ML}^{\gamma,\nu}) :=& n\left\{V_{\theta,\gamma,\nu}(\hat{U}^*) - V_{\theta}(U_{\gamma,\nu})\right\} \\ =& \frac{1}{\lambda_2}\left[\{u(\gamma) - \lambda_1\}^2 + \{u(\nu) - \lambda_1\}^2\right] + o(1), \\ d_n(\hat{\theta}_{MCL}, \hat{\theta}_{ML}^{\gamma,\nu}) :=& n\left\{V_{\theta,\gamma,\nu}(\tilde{U}_0) - V_{\theta}(U_{\gamma,\nu})\right\} \\ =& \frac{1}{\lambda_2}\left[\{u(\gamma) - \lambda_1\}^2 + \{u(\nu) - \lambda_1\}^2\right] + o(1) \end{aligned}$$

as $n \to \infty$, respectively.

The proof is straightforward from Theorem 3.3.1, 3.4.1, and 3.5.1.

Remark 3.6.1 It is seen from (1.6) and Theorem 3.6.1 that the ratio of the asymptotic variance of $\hat{U}^*$ to that of $U_{\gamma,\nu}$ is given by

$$R_n(\hat{\theta}_{ML^*}, \hat{\theta}_{ML}^{\gamma,\nu}) = 1 + \frac{1}{\lambda_2 n}\left[\{u(\gamma) - \lambda_1\}^2 + \{u(\nu) - \lambda_1\}^2\right] + o\left(\frac{1}{n}\right),$$

and similarly from (1.6) and Theorem 3.6.1

$$R_n(\hat{\theta}_{MCL}, \hat{\theta}_{ML}^{\gamma,\nu}) = 1 + \frac{1}{\lambda_2 n}\left[\{u(\gamma) - \lambda_1\}^2 + \{u(\nu) - \lambda_1\}^2\right] + o\left(\frac{1}{n}\right).$$

From the consideration of models in Sect. 1.1, using (1.5), (1.6), and Theorem 3.6.1 we see that the difference between the asymptotic models $M(\hat{\theta}_{ML^*}, \gamma, \nu)$ and $M(\hat{\theta}_{ML}^{\gamma,\nu}, \gamma, \nu)$ is given by $d_n(\hat{\theta}_{ML^*}, \hat{\theta}_{ML}^{\gamma,\nu})$ or $R_n(\hat{\theta}_{ML^*}, \hat{\theta}_{ML}^{\gamma,\nu})$ up to the second order, through the MLE of θ. In a similar way to the above, the difference between the asymptotic models $M(\hat{\theta}_{MCL}, \gamma, \nu)$ and $M(\hat{\theta}_{MCL}^{\gamma,\nu}, \gamma, \nu)$ is given by $d_n(\hat{\theta}_{MCL}, \hat{\theta}_{ML}^{\gamma,\nu})$ or $R_n(\hat{\theta}_{MCL}, \hat{\theta}_{ML}^{\gamma,\nu})$ up to the second order.

3.7 Examples

For a two-sided truncated exponential, a two-sided truncated normal, an upper-truncated Pareto, a two-sided truncated beta, and a two-sided truncated Erlang-type cases, the second-order asymptotic losses of the estimators are given as examples.

Example 3.7.1 **(Two-sided truncated exponential distribution)** Let $c = -\infty$, $d = \infty$, $a(x) \equiv 1$ and $u(x) = -x$ for $-\infty < \gamma \leq x \leq \nu < \infty$ in the density (1.9). Since $b(\theta, \gamma, \nu) = (e^{-\theta\gamma} - e^{-\theta\nu})/\theta$ for $\theta \in \Theta = (0, \infty)$, it follows from (3.1) and Theorem 3.4.1 that

$$\lambda_1 = \frac{\partial}{\partial\theta} \log b(\theta, \gamma, \nu) = \frac{-\gamma e^{-\theta\gamma} + \nu e^{-\theta\nu}}{e^{-\theta\gamma} - e^{-\theta\nu}} - \frac{1}{\theta},$$

$$\lambda_2 = \frac{\partial^2}{\partial\theta^2} \log b(\theta, \gamma, \nu) = \frac{\gamma^2 e^{-\theta\gamma} - \nu^2 e^{-\theta\nu}}{e^{-\theta\gamma} - e^{-\theta\nu}} - \frac{(\gamma e^{-\theta\gamma} - \nu e^{-\theta\nu})^2}{(e^{-\theta\gamma} - e^{-\theta\nu})^2} + \frac{1}{\theta^2},$$

$$k(\theta, \gamma, \nu) = \frac{a(\gamma) e^{\theta u(\gamma)}}{b(\theta, \gamma, \nu)} = \frac{\theta e^{-\theta\gamma}}{e^{-\theta\gamma} - e^{-\theta\nu}},$$

$$\tilde{k}(\theta, \gamma, \nu) = \frac{a(\nu) e^{\theta u(\nu)}}{b(\theta, \gamma, \nu)} = \frac{\theta e^{-\theta\nu}}{e^{-\theta\gamma} - e^{-\theta\nu}}.$$

Then, it follows from (3.2), (3.4), and (3.12) that the solutions of θ of the following equations

$$\frac{\gamma e^{-\theta\gamma} - \nu e^{-\theta\nu}}{e^{-\theta\gamma} - e^{-\theta\nu}} + \frac{1}{\theta} = \bar{X},$$

$$\frac{X_{(1)} e^{-\theta X_{(1)}} - X_{(n)} e^{-\theta X_{(n)}}}{e^{-\theta X_{(1)}} - e^{-\theta X_{(n)}}} + \frac{1}{\theta} = \bar{X},$$

$$\frac{X_{(1)} e^{-\theta X_{(1)}} - X_{(n)} e^{-\theta X_{(n)}}}{e^{-\theta X_{(1)}} - e^{-\theta X_{(n)}}} + \frac{1}{\theta} = \frac{1}{n-2} \sum_{i=2}^{n-1} X_{(i)}$$

become $\hat{\theta}_{ML}^{\gamma,\nu}$, $\hat{\theta}_{ML}$, and $\hat{\theta}_{MCL}$, respectively, where $\bar{X} = (1/n) \sum_{i=1}^{n} X_i$. From (3.5), the bias-adjusted MLE is seen to be given by

$$\hat{\theta}_{ML^*} = \hat{\theta}_{ML} + \frac{1}{\hat{\lambda}_2 n} \left\{ \frac{1}{\hat{k}} \left(\frac{\partial \hat{\lambda}_1}{\partial \gamma} \right) - \frac{1}{\hat{\tilde{k}}} \left(\frac{\partial \hat{\lambda}_1}{\partial \nu} \right) \right\}$$

where $\hat{\lambda}_i = \lambda_i(\hat{\theta}_{ML}, X_{(1)}, X_{(n)})$ $(i = 1, 2)$, $\hat{k} = k(\hat{\theta}_{ML}, X_{(1)}, X_{(n)})$, $\hat{\tilde{k}} = \tilde{k}(\hat{\theta}_{ML}, X_{(1)}, X_{(n)})$, and

$$\frac{\partial \hat{\lambda}_1}{\partial \gamma} = \frac{\partial \lambda_1}{\partial \gamma}(\hat{\theta}_{ML}, X_{(1)}, X_{(n)}), \quad \frac{\partial \hat{\lambda}_1}{\partial \nu} = \frac{\partial \lambda_1}{\partial \nu}(\hat{\theta}_{ML}, X_{(1)}, X_{(n)})$$

Table 3.1 Values of $d_n(\hat{\theta}_{ML^*}, \hat{\theta}_{ML}^{\gamma,\nu})$ and $R_n(\hat{\theta}_{ML^*}, \hat{\theta}_{ML}^{\gamma,\nu})$ for $\theta = \gamma = 1$

ν	$d_n(\hat{\theta}_{ML^*}, \hat{\theta}_{ML}^{\gamma,\nu})$	$R_n(\hat{\theta}_{ML^*}, \hat{\theta}_{ML}^{\gamma,\nu})$
2	$6.4725 + o(1)$	$1 + \dfrac{6.4725}{n} + o\left(\dfrac{1}{n}\right)$
3	$7.9582 + o(1)$	$1 + \dfrac{7.9582}{n} + o\left(\dfrac{1}{n}\right)$
5	$14.8146 + o(1)$	$1 + \dfrac{14.8146}{n} + o\left(\dfrac{1}{n}\right)$

with

$$\frac{\partial \lambda_1}{\partial \gamma} = -\frac{e^{-\theta\gamma}}{e^{-\theta\gamma} - e^{-\theta\nu}} - \frac{\theta(\gamma - \nu)e^{-\theta(\gamma+\nu)}}{(e^{-\theta\gamma} - e^{-\theta\nu})^2},$$
$$\frac{\partial \lambda_1}{\partial \nu} = \frac{e^{-\theta\nu}}{e^{-\theta\gamma} - e^{-\theta\nu}} + \frac{\theta(\gamma - \nu)e^{-\theta(\gamma+\nu)}}{(e^{-\theta\gamma} - e^{-\theta\nu})^2}.$$

From Theorem 3.6.1, we obtain the second-order asymptotic losses

$$d_n(\hat{\theta}_{ML^*}, \hat{\theta}_{MCL}) = o(1),$$
$$d_n(\hat{\theta}_{ML^*}, \hat{\theta}_{ML}^{\gamma,\nu}) = d_n(\hat{\theta}_{MCL}, \hat{\theta}_{ML}^{\gamma,\nu}) = \frac{\left(\frac{1}{\theta} + \frac{(\gamma-\nu)e^{-\theta\gamma}}{e^{-\theta\gamma}-e^{-\theta\nu}}\right)^2 + \left(\frac{1}{\theta} + \frac{(\gamma-\nu)e^{-\theta\nu}}{e^{-\theta\gamma}-e^{-\theta\nu}}\right)^2}{\frac{1}{\theta^2} - \frac{(\gamma-\nu)^2 e^{-\theta(\gamma+\nu)}}{(e^{-\theta\gamma}-e^{-\theta\nu})^2}} + o(1)$$

as $n \to \infty$.

When $\theta = \gamma = 1$, and $\nu = 2, 3, 5$, the values of the second-order asymptotic loss $d_n(\hat{\theta}_{ML^*}, \hat{\theta}_{ML}^{\gamma,\nu})$ and the ratio $R_n(\hat{\theta}_{ML^*}, \hat{\theta}_{ML}^{\gamma,\nu})$ are obtained from the above and Remark 3.6.1 (see Table 3.1 and Fig. 3.1).

Example 3.7.2 **(Two-sided truncated normal distribution)** Let $c = -\infty, d = \infty$, $a(x) = e^{-x^2/2}$, and $u(x) = x$ for $-\infty < \gamma \le x \le \nu < \infty$ in the density (1.9). Since

$$b(\theta, \gamma, \nu) = \{\Phi(\theta - \gamma) - \Phi(\theta - \nu)\}/\phi(\theta)$$

for $\theta \in \Theta = (-\infty, \infty)$, it follows that

$$\lambda_1(\theta, \gamma, \nu) = \theta + \eta_{\gamma-\nu}(\theta - \gamma) + \eta_{\nu-\gamma}(\theta - \nu),$$
$$\lambda_2(\theta, \gamma, \nu) = 1 - (\theta - \gamma)\eta_{\gamma-\nu}(\theta - \gamma) - (\theta - \nu)\eta_{\nu-\gamma}(\theta - \nu) - \{\eta_{\gamma-\nu}(\theta - \gamma) + \eta_{\nu-\gamma}(\theta - \nu)\}^2,$$
$$k(\theta, \gamma, \nu) = \eta_{\gamma-\nu}(\theta - \gamma), \quad \tilde{k}(\theta, \gamma, \nu) = -\eta_{\nu-\gamma}(\theta - \nu)$$

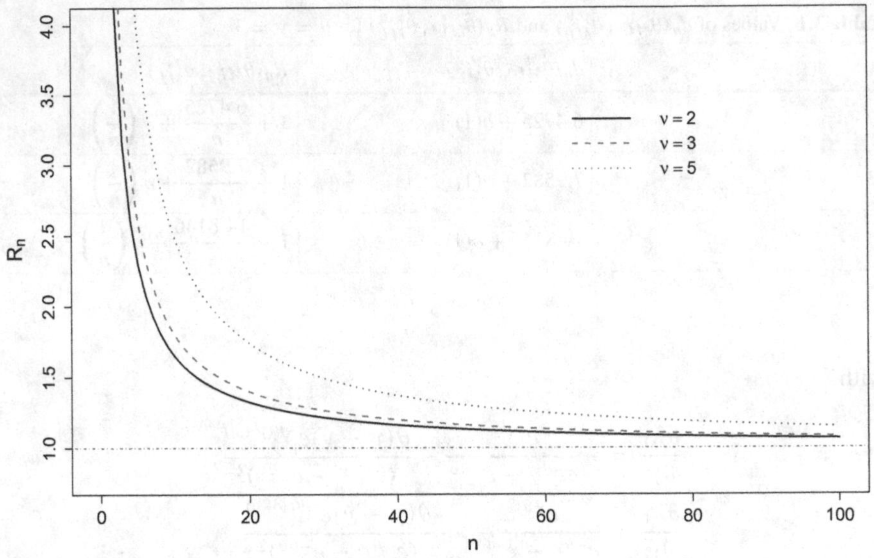

Fig. 3.1 Graph of the ratio $R_n(\hat{\theta}_{ML^*}, \hat{\theta}_{ML}^{\gamma,\nu})$ up to the order $1/n$ for $\theta = \gamma = 1$

where $\eta_\alpha(t) := \phi(t)/\{\Phi(t) - \Phi(t+\alpha)\}$ with $\Phi(t) = \int_{-\infty}^{t} \phi(x)dx$ and $\phi(x) = (1/\sqrt{2\pi})e^{-x^2/2}$ for $-\infty < x < \infty$. Then, it follows from (3.2), (3.4) and (3.12) that the solutions of the following equations

$$\theta + \eta_{\gamma-\nu}(\theta - \gamma) - \eta_{\nu-\gamma}(\theta - \nu) = \bar{X},$$
$$\theta + \eta_{X_{(1)}-X_{(n)}}(\theta - X_{(1)}) - \eta_{X_{(n)}-X_{(1)}}(\theta - X_{(n)}) = \bar{X}$$

and

$$\theta - \eta_{X_{(1)}-X_{(n)}}(\theta - X_{(1)}) - \eta_{X_{(n)}-X_{(1)}}(\theta - X_{(n)}) = \frac{1}{n-2}\sum_{i=2}^{n-1} X_{(i)}$$

become $\hat{\theta}_{ML}^{\gamma,\nu}$, $\hat{\theta}_{ML}$, and $\hat{\theta}_{MCL}$, respectively. From (3.5), the bias-adjusted MLE is seen to be given by

$$\hat{\theta}_{ML^*} = \hat{\theta}_{ML} + \frac{1}{\hat{\lambda}_2 n}\left\{\frac{1}{\hat{k}}\left(\frac{\partial\hat{\lambda}_1}{\partial\gamma}\right) - \frac{1}{\hat{\tilde{k}}}\left(\frac{\partial\hat{\lambda}_1}{\partial\nu}\right)\right\}$$

where

$$\hat{\lambda}_i = \lambda_i(\hat{\theta}_{ML}, X_{(1)}, X_{(n)})\ (i = 1, 2), \quad \hat{k} = k(\hat{\theta}_{ML}, X_{(1)}, X_{(n)}), \quad \hat{\tilde{k}} = \tilde{k}(\hat{\theta}_{ML}, X_{(1)}, X_{(n)})$$

Table 3.2 Values of $d_n(\hat{\theta}_{ML^*}, \hat{\theta}_{ML}^{\gamma,\nu})$ and $R_n(\hat{\theta}_{ML^*}, \hat{\theta}_{ML}^{\gamma,\nu})$ for $\theta = \gamma = 0$

ν	$d_n(\hat{\theta}_{ML^*}, \hat{\theta}_{ML}^{\gamma,\nu})$	$R_n(\hat{\theta}_{ML^*}, \hat{\theta}_{ML}^{\gamma,\nu})$
1	$6.3154 + o(1)$	$1 + \dfrac{6.3154}{n} + o\left(\dfrac{1}{n}\right)$
2	$8.5681 + o(1)$	$1 + \dfrac{8.5681}{n} + o\left(\dfrac{1}{n}\right)$
3	$15.8437 + o(1)$	$1 + \dfrac{15.8437}{n} + o\left(\dfrac{1}{n}\right)$

and

$$\frac{\partial \hat{\lambda}_1}{\partial \gamma} = \frac{\partial \lambda_1}{\partial \gamma}(\hat{\theta}_{ML}, X_{(1)}, X_{(n)}), \quad \frac{\partial \hat{\lambda}_1}{\partial \nu} = \frac{\partial \lambda_1}{\partial \nu}(\hat{\theta}_{ML}, X_{(1)}, X_{(n)})$$

with

$$\frac{\partial \lambda_1}{\partial \gamma} = \eta_{\gamma-\nu}(\theta - \gamma)\left\{\theta - \gamma - \eta_{\gamma-\nu}(\theta - \gamma) - \eta_{\nu-\gamma}(\theta - \nu)\right\},$$
$$\frac{\partial \lambda_1}{\partial \nu} = \eta_{\nu-\gamma}(\theta - \nu)\left\{\theta - \nu + \eta_{\gamma-\nu}(\theta - \gamma) + \eta_{\nu-\gamma}(\theta - \nu)\right\}.$$

From Theorem 3.6.1, we obtain the second-order asymptotic losses

$$\begin{aligned} &d_n(\hat{\theta}_{ML^*}, \hat{\theta}_{MCL}) = o(1), \\ &d_n(\hat{\theta}_{ML^*}, \hat{\theta}_{ML}^{\gamma,\nu}) = d_n(\hat{\theta}_{MCL}, \hat{\theta}_{ML}^{\gamma,\nu}) \\ &= \frac{\{\theta - \gamma + \eta_{\gamma-\nu}(\theta - \gamma) + \eta_{\nu-\gamma}(\theta - \nu)\}^2 + \{\theta - \nu + \eta_{\gamma-\nu}(\theta - \gamma) + \eta_{\nu-\gamma}(\theta - \nu)\}^2}{1 - (\theta - \gamma)\eta_{\gamma-\nu}(\theta - \gamma) - (\theta - \nu)\eta_{\nu-\gamma}(\theta - \nu) - \{\eta_{\gamma-\nu}(\theta - \gamma) + \eta_{\nu-\gamma}(\theta - \nu)\}^2} \\ &+ o(1) \end{aligned}$$

as $n \to \infty$.

When $\theta = \gamma = 0$, and $\nu = 1, 2, 3$, the values of the second-order asymptotic loss $d_n(\hat{\theta}_{ML^*}, \hat{\theta}_{ML}^{\gamma,\nu})$ of $\hat{\theta}_{ML^*}$ and the ratio $R_n(\hat{\theta}_{ML^*}, \hat{\theta}_{ML}^{\gamma,\nu})$ are obtained from the above and Remark 3.6.1 (see Table 3.2 and Fig. 3.2).

Example 3.7.3 **(Upper-truncated Pareto distribution)** For the Pareto distribution with an index parameter α to be estimated and two truncation parameters γ and ν as nuisance parameters, Aban et al. (2006) show the asymptotic normality of the MLEs $\tilde{\alpha}$ and $\hat{\alpha}$ of α in the case when γ and ν are known and the case when γ and ν are unknown, respectively. Although it is noted in Remark 2 of their paper that the asymptotic variance of $\hat{\alpha}$ is not the same as that of $\tilde{\alpha}$, it is seen from Theorems 3.3.1 and 3.4.1 that $\tilde{\alpha}$ and $\hat{\alpha}$ have the same asymptotic variance in the first order. However, in the second-order asymptotic comparison, a bias-adjustment of $\hat{\alpha}$ is needed and its second-order

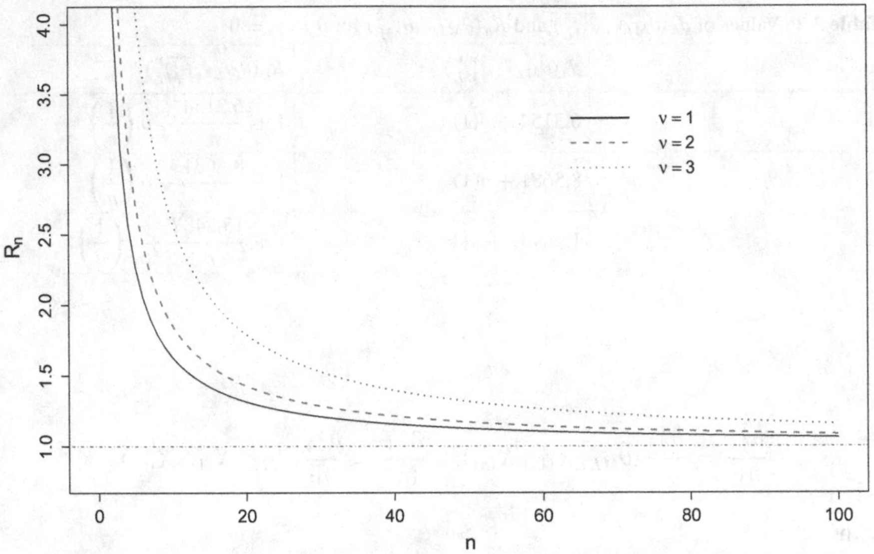

Fig. 3.2 Graph of the ratio $R_n(\hat{\theta}_{ML^*}, \hat{\theta}_{ML}^{\gamma,\nu})$ up to the order $1/n$ for $\theta = \gamma = 0$

asymptotic variance is different from that of $\tilde{\alpha}$, as below. Note that α is represented as θ in this chapter. Let $c = 0$, $d = \infty$, $a(x) = 1/x$, and $u(x) = -\log x$ for $0 < \gamma \le x \le \nu < \infty$ in the density (1.9). Then, $b(\theta, \gamma, \nu) = \{1 - (\gamma/\nu)^\theta\}/(\theta\gamma^\theta)$ for $\theta \in \Theta = (0, \infty)$. Letting $t = \log x$, $\gamma_0 = \log \gamma$, and $\nu_0 = \log \nu$, we see that (1.9) becomes

$$f(t; \theta, \gamma_0, \nu_0) = \begin{cases} \frac{\theta e^{\theta\gamma_0}}{1-e^{-\theta(\nu_0-\gamma_0)}} e^{-\theta t} & \text{for } \gamma_0 \le t \le \nu_0, \\ 0 & \text{otherwise.} \end{cases}$$

Hence, the upper-truncated Pareto distribution case is reduced to the two-sided truncated exponential one in Example 3.7.1. Replacing $\bar{X}$ and $X_{(i)}$ $(i = 1, \cdots, n)$ by $\overline{\log X} := (1/n)\sum_{i=1}^n \log X_i$ and $\log X_{(i)}$ $(i = 1, \cdots, n)$, respectively, in Example 3.7.1, we have the second-order asymptotic losses

$$\begin{aligned} & d_n(\hat{\theta}_{ML^*}, \hat{\theta}_{MCL}) = o(1), \\ & d_n(\hat{\theta}_{ML^*}, \hat{\theta}_{ML}^{\gamma,\nu}) = d_n(\hat{\theta}_{MCL}, \hat{\theta}_{ML}^{\gamma,\nu}) \\ & = \left\{\left(1 + \frac{\xi \log \xi}{1-\xi}\right)^2 + \left(1 + \frac{\log \xi}{1-\xi}\right)^2\right\} \Big/ \left\{1 - \frac{\xi(\log \xi)^2}{(1-\xi)^2}\right\} + o(1) \end{aligned}$$

as $n \to \infty$ where $\xi := (\gamma/\nu)^\theta$.

When $\theta = 0.8$, $\gamma = 1$, and $\nu = 5, 10, 15$, the values of the second-order asymptotic loss $d_n(\hat{\theta}_{ML^*}, \hat{\theta}_{ML}^{\gamma,\nu})$ of $\hat{\theta}_{ML^*}$ and the ratio $R_n(\hat{\theta}_{ML^*}, \hat{\theta}_{ML}^{\gamma,\nu})$ are obtained from the

Table 3.3 Values of $d_n(\hat{\theta}_{ML^*}, \hat{\theta}_{ML}^{\gamma,\nu})$ and $R_n(\hat{\theta}_{ML^*}, \hat{\theta}_{ML}^{\gamma,\nu})$ for $\theta = 0.8$ and $\gamma = 1$

ν	$d_n(\hat{\theta}_{ML^*}, \hat{\theta}_{ML}^{\gamma,\nu})$	$R_n(\hat{\theta}_{ML^*}, \hat{\theta}_{ML}^{\gamma,\nu})$
5	$6.7898 + o(1)$	$1 + \dfrac{6.7898}{n} + o\left(\dfrac{1}{n}\right)$
10	$7.6495 + o(1)$	$1 + \dfrac{7.6495}{n} + o\left(\dfrac{1}{n}\right)$
15	$8.3155 + o(1)$	$1 + \dfrac{8.3155}{n} + o\left(\dfrac{1}{n}\right)$

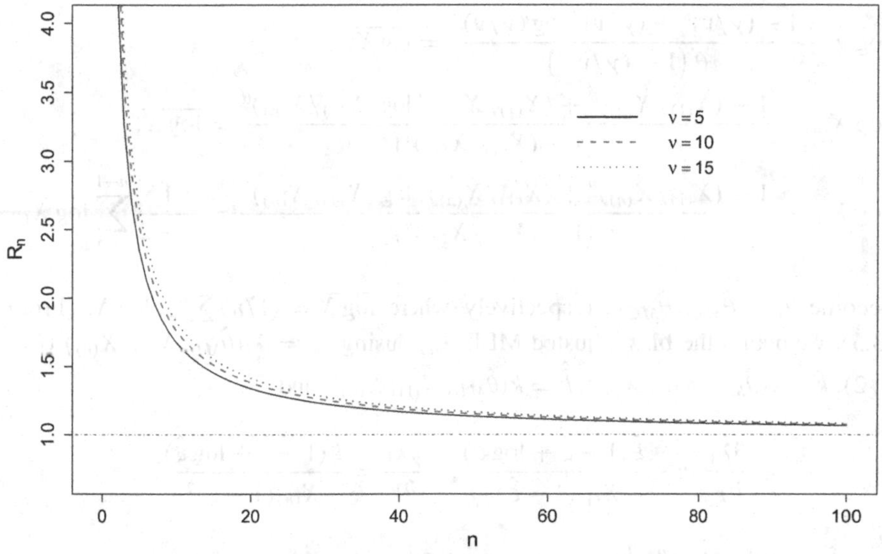

Fig. 3.3 Graph of the ratio $R_n(\hat{\theta}_{ML^*}, \hat{\theta}_{ML}^{\gamma,\nu})$ up to the order $1/n$ for $\theta = 0.8$ and $\gamma = 1$

above and Remark 3.6.1 (see Table 3.3 and Fig. 3.3). In Aban et al. (2006), the performance of the MLE is compared with that of the estimators of Hill (1975) and Beg (1981) when $\theta = 0.8$, $\gamma = 1$, and $\nu = 10$.

Example 3.7.4 **(Two-sided truncated beta distribution)** Let $c = 0$, $d = 1$, $a(x) = x^{-1}$, and $u(x) = \log x$ for $0 < \gamma \le x \le \nu < 1$ in the density (1.9). Note that the density is uniform over the interval $[\gamma, \nu]$ when $\theta = 1$. Since $b(\theta, \gamma, \nu) = \theta^{-1}\nu^{\theta}(1 - (\gamma/\nu)^{\theta})$ for $\theta \in \Theta = (0, \infty)$, it follows from (3.1) and Theorem 3.4.1 that

$$\lambda_1(\theta, \gamma, \nu) = \log \nu - \frac{1 - \xi + \xi \log \xi}{\theta(1-\xi)},$$
$$\lambda_2(\theta, \gamma, \nu) = \frac{1}{\theta^2} - \frac{\xi}{\theta^2(1-\xi)^2}(\log \xi)^2,$$
$$k(\theta, \gamma, \nu) = \frac{\theta\xi}{\gamma(1-\xi)}, \quad \tilde{k}(\theta, \gamma, \nu) = \frac{\theta}{\nu(1-\xi)}$$

where $\xi = (\gamma/\nu)^\theta$. Then, it follows from (3.2), (3.4), and (3.12) that the solutions of θ of the following

$$\log \nu - \frac{1 - (\gamma/\nu)^\theta + (\gamma/\nu)^\theta \log (\gamma/\nu)^\theta}{\theta\left(1 - (\gamma/\nu)^\theta\right)} = \overline{\log X},$$
$$\log X_{(n)} - \frac{1 - (X_{(1)}/X_{(n)})^\theta + (X_{(1)}/X_{(n)})^\theta \log (X_{(1)}/X_{(n)})^\theta}{\theta\left(1 - (X_{(1)}/X_{(n)})^\theta\right)} = \overline{\log X},$$
$$\log X_{(n)} - \frac{1 - (X_{(1)}/X_{(n)})^\theta + (X_{(1)}/X_{(n)})^\theta \log (X_{(1)}/X_{(n)})^\theta}{\theta\left(1 - (X_{(1)}/X_{(n)})^\theta\right)} = \frac{1}{n-2}\sum_{i=2}^{n-1} \log X_{(i)}$$

become $\hat{\theta}_{ML}^{\gamma,\nu}$, $\hat{\theta}_{ML}$, $\hat{\theta}_{MCL}$, respectively where $\overline{\log X} = (1/n)\sum_{i=1}^n \log X_i$. From (3.5), we obtain the bias-adjusted MLE $\hat{\theta}_{ML^*}$ using $\hat{\lambda}_i := \lambda_i(\hat{\theta}_{ML}, X_{(1)}, X_{(n)})$ $(i = 1, 2)$, $\hat{k} = k(\hat{\theta}_{ML}, X_{(1)}, X_{(n)})$, $\hat{\tilde{k}} = \tilde{k}(\hat{\theta}_{ML}, X_{(1)}, X_{(n)})$, and

$$\frac{\partial \hat{\lambda}_1}{\partial \gamma} = -\frac{\hat{\xi}(1 - \hat{\xi} + \log \hat{\xi})}{X_{(1)}(1-\hat{\xi})^2}, \quad \frac{\partial \hat{\lambda}_1}{\partial \nu} = \frac{\hat{\xi}(1 - \hat{\xi} + \log \hat{\xi})}{X_{(n)}(1-\hat{\xi})^2},$$

with $\hat{\xi} = (X_{(1)}/X_{(n)})^{\hat{\theta}_{ML}}$. From Theorem 3.6.1, we have the second-order asymptotic losses

$$d_n(\hat{\theta}_{ML^*}, \hat{\theta}_{MCL}) = o(1),$$
$$d_n(\hat{\theta}_{ML^*}, \hat{\theta}_{ML}^{\gamma,\nu}) = d_n(\hat{\theta}_{MCL}, \hat{\theta}_{ML}^{\gamma,\nu}) = \frac{(1 - \xi - \log \xi)^2 + (1 - \xi + \xi \log \xi)^2}{(1-\xi)^2 - \xi(\log \xi)^2} + o(1),$$

as $n \to \infty$.

When $\theta = 3$, $\gamma = 1/4$, and $\nu = 1/2,\ 2/3,\ 5/6$, the values of the second-order asymptotic loss $d_n(\hat{\theta}_{ML^*}, \hat{\theta}_{ML}^{\gamma,\nu})$ and the ratio $R_n(\hat{\theta}_{ML^*}, \hat{\theta}_{ML}^{\gamma,\nu})$ are obtained from the above and Remark 3.6.1 (see Table 3.4 and Fig. 3.4).

Example 3.7.5 **(Two-sided truncated Erlang type distribution)** Let $c = -\infty$, $d = \infty$, $a(x) = |x|^{j-1}$, and $u(x) = -|x|$ for $-\infty < \gamma \le x \le \nu < \infty$ in the density (1.9) where $j = 1, 2, \ldots$. Then, for each $j = 1, 2, \ldots$, $b_j(\theta, \gamma, \nu) = \int_\gamma^\nu |x|^{j-1} e^{-\theta|x|} dx$ for $\theta \in \Theta = (0, \infty)$. In particular, the distribution is a two-sided truncated bilateral

Table 3.4 Values of $d_n(\hat{\theta}_{ML^*}, \hat{\theta}_{ML}^{\gamma,\nu})$ and $R_n(\hat{\theta}_{ML^*}, \hat{\theta}_{ML}^{\gamma,\nu})$ for $\theta = 3$ and $\gamma = 1/4$

ν	$d_n(\hat{\theta}_{ML^*}, \hat{\theta}_{ML}^{\gamma,\nu})$	$R_n(\hat{\theta}_{ML^*}, \hat{\theta}_{ML}^{\gamma,\nu})$
1/2	$8.1247 + o(1)$	$1 + \dfrac{8.1247}{n} + o\left(\dfrac{1}{n}\right)$
2/3	$10.4562 + o(1)$	$1 + \dfrac{10.4562}{n} + o\left(\dfrac{1}{n}\right)$
5/6	$13.0034 + o(1)$	$1 + \dfrac{13.0034}{n} + o\left(\dfrac{1}{n}\right)$

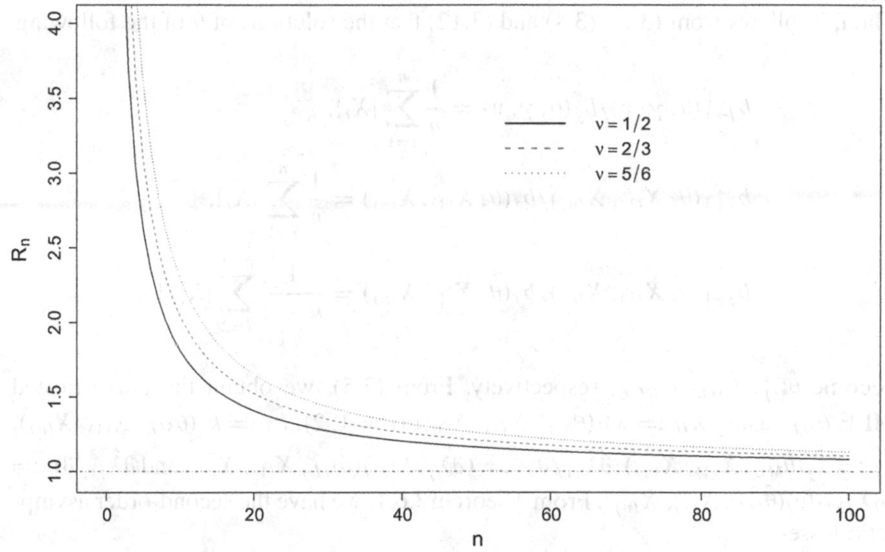

Fig. 3.4 Graph of the ratio $R_n(\hat{\theta}_{ML^*}, \hat{\theta}_{ML}^{\gamma,\nu})$ up to the order $1/n$ for $\theta = 3$ and $\gamma = 1/4$

exponential (Laplace) distribution when $j = 1$. If $\gamma > 0$, then the distribution is a two-sided truncated Erlang distribution with density

$$f_j(x; \theta, \gamma, \nu) = \begin{cases} \frac{1}{b(\theta,\gamma,\nu)} x^{j-1} e^{-\theta x} & \text{for } 0 < \gamma \le x \le \nu < \infty, \\ 0 & \text{otherwise} \end{cases}$$

where $j = 1, 2, \ldots$, and $b(\theta, \gamma, \nu) = \int_\gamma^\nu x^{j-1} e^{-\theta x} dx$ for $\theta > 0$. If $\nu < 0$, then the distribution becomes the above two-sided truncated Erlang distribution by the change of variable since $-\infty < \gamma < \nu < 0$. Hence, we consider the case when $-\infty < \gamma \le 0 < \nu < \infty$. Let j be arbitrarily fixed in $\{1, 2, \ldots\}$. Since $\partial b_j / \partial \theta = -b_{j+1}$, it follows from (3.1) and Theorem 3.4.1 that

$$\lambda_{j1}(\theta, \gamma, \nu) = \frac{\partial}{\partial\theta}\log b_j(\theta, \gamma, \nu) = -\frac{b_{j+1}}{b_j},$$
$$\lambda_{j2}(\theta, \gamma, \nu) = \frac{\partial^2}{\partial\theta^2}\log b_j(\theta, \gamma, \nu) = \frac{b_{j+2}}{b_j} - \left(\frac{b_{j+1}}{b_j}\right)^2,$$
$$k_j(\theta, \gamma, \nu) = \frac{a(\gamma)e^{\theta u(\gamma)}}{b_j(\theta, \gamma, \nu)} = \frac{|\gamma|^{j-1}e^{-\theta|\gamma|}}{b_j(\theta, \gamma, \nu)},$$
$$\tilde{k}_j(\theta, \gamma, \nu) = \frac{a(\nu)e^{\theta u(\nu)}}{b_j(\theta, \gamma, \nu)} = \frac{|\nu|^{j-1}e^{-\theta|\nu|}}{b_j(\theta, \gamma, \nu)}.$$

Then, it follows from (3.2), (3.4) and (3.12) that the solutions of θ of the following

$$b_{j+1}(\theta, \gamma, \nu)/b_j(\theta, \gamma, \nu) = \frac{1}{n}\sum_{i=1}^{n}|X_i|,$$

$$b_{j+1}(\theta, X_{(1)}, X_{(n)})/b_j(\theta, X_{(1)}, X_{(n)}) = \frac{1}{n}\sum_{i=1}^{n}|X_i|,$$

$$b_{j+1}(\theta, X_{(1)}, X_{(n)})/b_j(\theta, X_{(1)}, X_{(n)}) = \frac{1}{n-2}\sum_{i=2}^{n-1}|X_{(i)}|$$

become $\hat{\theta}_{ML}^{\gamma,\nu}$, $\hat{\theta}_{ML}$, $\hat{\theta}_{MCL}$, respectively. From (3.5), we obtain the bias-adjusted MLE $\hat{\theta}_{ML^*}$ using $\hat{\lambda}_{ji} := \lambda_{ji}(\hat{\theta}_{ML}, X_{(1)}, X_{(n)})$ $(i = 1, 2)$, $\hat{k}_j := k_j(\hat{\theta}_{ML}, X_{(1)}, X_{(n)})$, $\hat{\tilde{k}}_j := \tilde{k}_j(\hat{\theta}_{ML}, X_{(1)}, X_{(n)})$, $\partial\hat{\lambda}_{j1}/\partial\gamma := (\partial\lambda_{j1}/\partial\gamma)(\hat{\theta}_{ML}, X_{(1)}, X_{(n)})$ and $\partial\hat{\lambda}_{j1}/\partial\nu := (\partial\lambda_{j1}/\partial\nu)(\hat{\theta}_{ML}, X_{(1)}, X_{(n)})$. From Theorem 3.6.1, we have the second-order asymptotic losses

$$d_n(\hat{\theta}_{ML^*}, \hat{\theta}_{MCL}) = o(1),$$
$$d_n(\hat{\theta}_{ML^*}, \hat{\theta}_{ML}^{\gamma,\nu}) = d_n(\hat{\theta}_{MCL}, \hat{\theta}_{ML}^{\gamma,\nu}) = \frac{(b_{j+1} - \gamma b_j)^2 + (b_{j+1} - \nu b_j)^2}{b_j b_{j+2} - b_{j+1}^2} + o(1),$$

as $n \to \infty$. If $j = 2$, then

$$b_2 = -\frac{1}{\theta}\left(\nu + \frac{1}{\theta}\right)e^{-\theta\nu} + \frac{1}{\theta}\left(\gamma + \frac{1}{\theta}\right)e^{-\theta\gamma},$$
$$b_3 = -\frac{1}{\theta}\left(\nu^2 + \frac{2}{\theta}\nu + \frac{2}{\theta^2}\right)e^{-\theta\nu} + \frac{1}{\theta}\left(\gamma^2 + \frac{2}{\theta}\gamma + \frac{2}{\theta^2}\right)e^{-\theta\gamma},$$
$$b_4 = -\frac{1}{\theta}\left(\nu^3 + \frac{3}{\theta}\nu^2 + \frac{6}{\theta^2}\nu + \frac{6}{\theta^3}\right)e^{-\theta\nu} + \frac{1}{\theta}\left(\gamma^3 + \frac{3}{\theta}\gamma^2 + \frac{6}{\theta^2}\gamma + \frac{6}{\theta^3}\right)e^{-\theta\gamma}.$$

Table 3.5 Values of $d_n(\hat{\theta}_{ML^*}, \hat{\theta}_{ML}^{0,\nu})$ and $R_n(\hat{\theta}_{ML^*}, \hat{\theta}_{ML}^{0,\nu})$ for $j = 2, \theta = 1$, and $\gamma = 0$

ν	$d_n(\hat{\theta}_{ML^*}, \hat{\theta}_{ML}^{0,\nu})$	$R_n(\hat{\theta}_{ML^*}, \hat{\theta}_{ML}^{0,\nu})$
1	$8.4737 + o(1)$	$1 + \dfrac{8.4737}{n} + o\left(\dfrac{1}{n}\right)$
2	$7.8107 + o(1)$	$1 + \dfrac{7.8107}{n} + o\left(\dfrac{1}{n}\right)$
3	$7.9360 + o(1)$	$1 + \dfrac{7.9360}{n} + o\left(\dfrac{1}{n}\right)$

Further, if $\theta = 1$ and $\gamma = 0$, then

$$b_2 = -(\nu + 1)e^{-\nu} + 1, \quad b_3 = -(\nu^2 + 2\nu + 2)e^{-\nu} + 2,$$
$$b_4 = -(\nu^3 + 3\nu^2 + 6\nu + 6)e^{-\nu} + 6,$$

hence

$$d_n(\hat{\theta}_{ML^*}, \hat{\theta}_{ML}^{0,\nu}) = \frac{(\nu^4 + 4\nu^3 + 9\nu^2 + 12\nu + 8)e^{-2\nu} - 2(\nu^2 + 4\nu + 8)e^{-\nu} + \nu^2 - 4\nu + 8}{(\nu^2 + 4\nu + 2)e^{-2\nu} - (\nu^3 - \nu^2 + 4\nu + 4)e^{-\nu} + 2} + o(1),$$

as $n \to \infty$. When $j = 2, \theta = 1, \gamma = 0$ and $\nu = 1, 2, 3$, the values of the second-order asymptotic loss $d_n(\hat{\theta}_{ML^*}, \hat{\theta}_{ML}^{\gamma,\nu})$ of $\hat{\theta}_{ML^*}$ and the ratio $R_n(\hat{\theta}_{ML^*}, \hat{\theta}_{ML}^{\gamma,\nu})$ are obtained from the above and Remark 3.6.1 (see Table 3.5 and Fig. 3.5). Note that letting $\theta = 0$ formally in the two-sided truncated Erlang-type distribution when $j = 1$, we get the uniform distribution over the interval $[\gamma, \nu]$.

Example 3.7.6 **(Two-sided truncated lognormal distribution)** Let $c = 0, d = \infty$, $a(x) = x^{-1}\exp\{-(1/2)(\log x)^2\}$, and $u(x) = \log x$ for $0 < \gamma \le x \le \nu < \infty$ in the density (1.9). Then,

$$b(\theta, \gamma, \nu) = \{\Phi(\theta - \log \gamma) - \Phi(\theta - \log \nu)\}/\phi(\theta)$$

for $\theta \in \Theta = (-\infty, \infty)$ where $\Phi(x) = \int_{-\infty}^{x} \phi(t)dt$ with $\phi(t) = (1/\sqrt{2\pi})e^{-t^2/2}$ for $-\infty < t < \infty$. Letting $t = \log x$, $\gamma_0 = \log \gamma$, and $\nu_0 = \log \nu$, we see that (1.9) becomes

$$f(t; \theta, \gamma_0, \nu_0) = \begin{cases} \frac{1}{\sqrt{2\pi}\{\Phi(\theta - \gamma_0) - \Phi(\theta - \nu_0)\}} e^{-(t-\theta)^2/2} & \text{for } -\infty < \gamma_0 \le t \le \nu_0 < \infty, \\ 0 & \text{otherwise.} \end{cases}$$

Hence, the two-sided truncated lognormal case is reduced to the two-sided truncated normal one in Example 3.7.2.

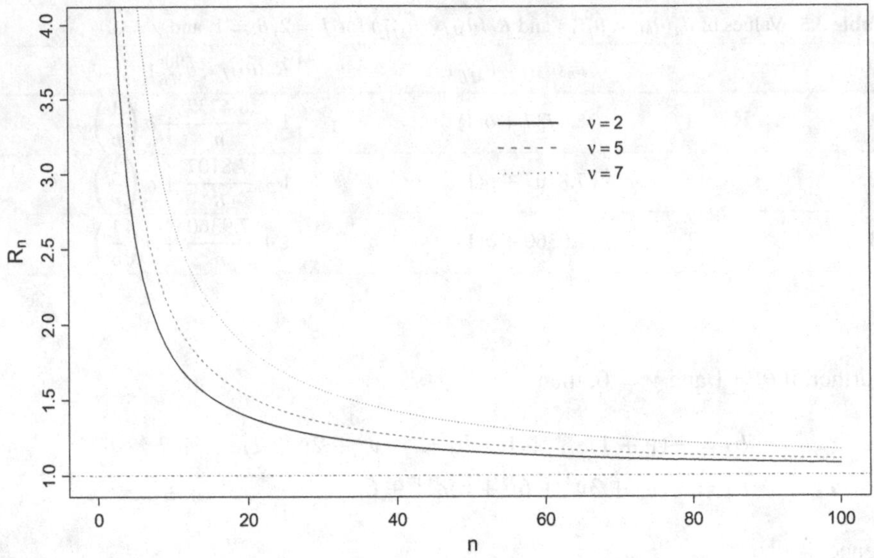

Fig. 3.5 Graph of the ratio $R_n(\hat{\theta}_{ML^*}, \hat{\theta}_{ML}^{0,\nu})$ up to the order $1/n$ for $j = 2, \theta = 1$ and $\gamma = 0$

3.8 Concluding Remarks

In this chapter, the corresponding results on a comparison of the estimators to the case of oTEF in Chap. 2 are obtained in the case of a tTEF of distributions with a natural parameter θ and two truncation parameters γ and ν as nuisance ones, including the upper-truncated Pareto distribution which is important in applications to finance, hydrology, and atmospheric science as is seen in Aban et al. (2006). In particular, the second-order asymptotic losses of $\hat{\theta}_{ML^*}$ and $\hat{\theta}_{MCL}$ given by Theorem 3.6.1 are seen to be quite simple, which results from a tTEF of distributions. Indeed, as is seen from the form (1.9) of density, the structure of the regular and non-regular parts of (1.9) reflects in that of the second-order asymptotic variances of $U_{\gamma,\nu} = \sqrt{\lambda_2 n}(\hat{\theta}_{ML}^{\gamma,\nu} - \theta)$, $\hat{U}^* = \sqrt{\lambda_2 n}(\hat{\theta}_{ML^*} - \theta)$ and $\tilde{U}_0 = \sqrt{\lambda_2 n}(\hat{\theta}_{MCL} - \theta)$ in Theorems 3.3.1, 3.4.1, and 3.5.1. The regular part corresponds to the term of order n^{-1} in the second-order asymptotic variance of $U_{\gamma,\nu}$ where γ and ν are known. When γ and ν are unknown, the second-order asymptotic variances of $\hat{U}^*$ and $\tilde{U}_0$ consist of the corresponding regular term and the non-regular one with the term depending on $u(\gamma)$ and $u(\nu)$ in the second order, i.e., the order n^{-1}. The results arise from giving full consideration to the typical non-regular case up to the second order. Furthermore, in a similar way to the above, the results may be extended to the case of a more general truncated family of distributions.

3.9 Appendix B1

Before proving Theorem 3.3.1, we prepare three lemmas (the proofs are given in Appendix B2 of the next section).

Lemma 3.9.1 *The second-order asymptotic densities of* $T_{(1)}$ *and* $T_{(n)}$ *are given by*

$$f_{T_{(1)}}(t) = k(\theta,\gamma,\nu)e^{-k(\theta,\gamma,\nu)t} - \frac{1}{2n}\left\{\frac{\partial}{\partial\gamma}\log k(\theta,\gamma,\nu)\right\}\left\{k(\theta,\gamma,\nu)t^2 - 2t\right\}k(\theta,\gamma,\nu)e^{-k(\theta,\gamma,\nu)t} + O\left(\frac{1}{n^2}\right) \tag{3.14}$$

for $t > 0$, *and*

$$f_{T_{(n)}}(t) = \tilde{k}(\theta,\gamma,\nu)e^{\tilde{k}(\theta,\gamma,\nu)t} + \frac{1}{2n}\left\{\frac{\partial}{\partial\nu}\log\tilde{k}(\theta,\gamma,\nu)\right\}\left\{\tilde{k}(\theta,\gamma,\nu)t^2 + 2t\right\}\tilde{k}(\theta,\gamma,\nu)e^{\tilde{k}(\theta,\gamma,\nu)t} + O\left(\frac{1}{n^2}\right) \tag{3.15}$$

for $t < 0$, *and*

$$E_{\theta,\gamma,\nu}(T_{(1)}) = \frac{1}{k(\theta,\gamma,\nu)} + \frac{A(\theta,\gamma,\nu)}{n} + O\left(\frac{1}{n^2}\right), \tag{3.16}$$

$$E_{\theta,\gamma,\nu}(T_{(n)}) = -\frac{1}{\tilde{k}(\theta,\gamma,\nu)} - \frac{\tilde{A}(\theta,\gamma,\nu)}{n} + O\left(\frac{1}{n^2}\right), \tag{3.17}$$

$$E_{\theta,\gamma,\nu}(T_{(1)}^2) = \frac{2}{k^2(\theta,\gamma,\nu)} + \frac{6A(\theta,\gamma,\nu)}{k(\theta,\gamma,\nu)n} + O\left(\frac{1}{n^2}\right), \tag{3.18}$$

$$E_{\theta,\gamma,\nu}(T_{(n)}^2) = \frac{2}{\tilde{k}^2(\theta,\gamma,\nu)} + \frac{6\tilde{A}(\theta,\gamma,\nu)}{\tilde{k}(\theta,\gamma,\nu)n} + O\left(\frac{1}{n^2}\right) \tag{3.19}$$

where

$$k(\theta,\gamma,\nu) = a(\gamma)e^{\theta u(\gamma)}/b(\theta,\gamma,\nu), \quad \tilde{k}(\theta,\gamma,\nu) = a(\nu)e^{\theta u(\nu)}/b(\theta,\gamma,\nu),$$

$$A(\theta,\gamma,\nu) := -\frac{1}{k^2(\theta,\gamma,\nu)}\frac{\partial}{\partial\gamma}\log k(\theta,\gamma,\nu),$$

$$\tilde{A}(\theta,\gamma,\nu) := \frac{1}{\tilde{k}^2(\theta,\gamma,\nu)}\frac{\partial}{\partial\nu}\log\tilde{k}(\theta,\gamma,\nu).$$

Lemma 3.9.2 *It holds that*

$$E_{\theta,\gamma,\nu}(Z_1 T_{(1)}) = \frac{1}{k\sqrt{\lambda_2 n}}\left\{u(\gamma) - \lambda_1 + \frac{2}{k}\left(\frac{\partial\lambda_1}{\partial\gamma}\right)\right\} + O\left(\frac{1}{n\sqrt{n}}\right), \tag{3.20}$$

$$E_{\theta,\gamma,\nu}(Z_1 T_{(n)}) = -\frac{1}{\tilde{k}\sqrt{\lambda_2 n}}\left\{u(\nu) - \lambda_1 - \frac{2}{\tilde{k}}\left(\frac{\partial\lambda_1}{\partial\nu}\right)\right\} + O\left(\frac{1}{n\sqrt{n}}\right) \tag{3.21}$$

where $k = k(\theta, \gamma, \nu)$ *and* $\lambda_i = \lambda_i(\theta, \gamma, \nu)$ $(i = 1, 2)$.

Lemma 3.9.3 *It holds that*

$$E_{\theta,\gamma,\nu}(Z_1^2 T_{(1)}) = \frac{1}{k} + O\left(\frac{1}{n}\right), \tag{3.22}$$

$$E_{\theta,\gamma,\nu}(Z_1^2 T_{(n)}) = -\frac{1}{\tilde{k}} + O\left(\frac{1}{n}\right) \tag{3.23}$$

where $k = k(\theta, \gamma, \nu)$ *and* $\tilde{k} = \tilde{k}(\theta, \gamma, \nu)$.

The proof of Theorem 3.4.1 Since, for $(\theta, \gamma, \nu) \in \Theta \times (c, X_{(1)}) \times (X_{(n)}, d)$

$$\begin{aligned}
&\lambda_1(\hat{\theta}_{ML}, X_{(1)}, X_{(n)}) \\
&= \lambda_1(\theta, \gamma, \nu) + \left\{\frac{\partial}{\partial\theta}\lambda_1(\theta, \gamma, \nu)\right\}(\hat{\theta}_{ML} - \theta) + \left\{\frac{\partial}{\partial\gamma}\lambda_1(\theta, \gamma, \nu)\right\}(X_{(1)} - \gamma) \\
&+ \left\{\frac{\partial}{\partial\nu}\lambda_1(\theta, \gamma, \nu)\right\}(X_{(n)} - \nu) + \frac{1}{2}\Bigg[\left\{\frac{\partial^2}{\partial\theta^2}\lambda_1(\theta, \gamma, \nu)\right\}(\hat{\theta}_{ML} - \theta)^2 \\
&+ 2\left\{\frac{\partial^2}{\partial\theta\partial\gamma}\lambda_1(\theta, \gamma, \nu)\right\}(\hat{\theta}_{ML} - \theta)(X_{(1)} - \gamma) \\
&+ 2\left\{\frac{\partial^2}{\partial\theta\partial\nu}\lambda_1(\theta, \gamma, \nu)\right\}(\hat{\theta}_{ML} - \theta)(X_{(n)} - \nu) \\
&+ \left\{\frac{\partial^2}{\partial\gamma^2}\lambda_1(\theta, \gamma, \nu)\right\}(X_{(1)} - \gamma)^2 + 2\left\{\frac{\partial^2}{\partial\gamma\partial\nu}\lambda_1(\theta, \gamma, \nu)\right\}(X_{(1)} - \gamma)(X_{(n)} - \nu) \\
&+ \left\{\frac{\partial^2}{\partial\nu^2}\lambda_1(\theta, \gamma, \nu)\right\}(X_{(n)} - \nu)^2\Bigg] + \frac{1}{6}\left\{\frac{\partial^3}{\partial\theta^3}\lambda_1(\theta, \gamma)\right\}(\hat{\theta}_{ML} - \theta)^3 + \cdots,
\end{aligned} \tag{3.24}$$

noting $\hat{U} = \sqrt{\lambda_2 n}(\hat{\theta}_{ML} - \theta)$, $T_{(1)} = n(X_{(1)} - \gamma)$, and $T_{(n)} = n(X_{(n)} - \nu)$, we have from (3.4) and (3.24)

$$
\begin{aligned}
0 =& \sqrt{\frac{\lambda_2}{n}} Z_1 - \sqrt{\frac{\lambda_2}{n}} \hat{U} - \frac{1}{n}\left(\frac{\partial \lambda_1}{\partial \gamma}\right) T_{(1)} - \frac{1}{n}\left(\frac{\partial \lambda_1}{\partial \nu}\right) T_{(n)} - \frac{\lambda_3}{2\lambda_2 n}\hat{U}^2 \\
& - \frac{1}{\sqrt{\lambda_2 n} n}\left(\frac{\partial \lambda_2}{\partial \gamma}\right) \hat{U} T_{(1)} - \frac{1}{\sqrt{\lambda_2 n} n}\left(\frac{\partial \lambda_2}{\partial \nu}\right) \hat{U} T_{(n)} - \frac{\lambda_4}{6\lambda_2^{3/2} n\sqrt{n}}\hat{U}^3 \\
& + O_p\left(\frac{1}{n^2}\right),
\end{aligned}
$$

which implies that the stochastic expansion of $\hat{U}$ is given by

$$
\begin{aligned}
\hat{U} =& Z_1 - \frac{\lambda_3}{2\lambda_2^{3/2}\sqrt{n}} Z_1^2 - \frac{1}{\sqrt{\lambda_2 n}}\left\{\left(\frac{\partial \lambda_1}{\partial \gamma}\right) T_{(1)} + \left(\frac{\partial \lambda_1}{\partial \nu}\right) T_{(n)}\right\} \\
& + \frac{1}{\lambda_2 n} Z_1 \left\{\delta_1 T_{(1)} + \delta_2 T_{(n)}\right\} + \frac{1}{2n}\left(\frac{\lambda_3^2}{\lambda_2^3} - \frac{\lambda_4}{3\lambda_2^2}\right) Z_1^3 + O_p\left(\frac{1}{n\sqrt{n}}\right)
\end{aligned}
\tag{3.25}
$$

where

$$
\delta_1 := \frac{\lambda_3}{\lambda_2}\left(\frac{\partial \lambda_1}{\partial \gamma}\right) - \frac{\partial \lambda_2}{\partial \gamma}, \quad \delta_2 := \frac{\lambda_3}{\lambda_2}\left(\frac{\partial \lambda_1}{\partial \nu}\right) - \frac{\partial \lambda_2}{\partial \nu}.
$$

Hence, we obtain (3.7). Here,

$$
E_{\theta,\gamma,\nu}(Z_1) = 0, \quad E_{\theta,\gamma,\nu}(Z_1^2) = 1, \quad E_{\theta,\gamma,\nu}(Z_1^3) = \frac{\lambda_3}{\lambda_2^{3/2}\sqrt{n}}, \quad E_{\theta,\gamma,\nu}(Z_1^4) = 3 + \frac{\lambda_4}{\lambda_2^2 n}. \tag{3.26}
$$

Then, it follows from (3.26) that

$$
\begin{aligned}
E_{\theta,\gamma,\nu}(\hat{U}) =& -\frac{\lambda_3}{2\lambda_2^{3/2}\sqrt{n}} - \frac{1}{\sqrt{\lambda_2 n}}\left\{\left(\frac{\partial \lambda_1}{\partial \gamma}\right) E_{\theta,\gamma,\nu}(T_{(1)}) + \left(\frac{\partial \lambda_1}{\partial \nu}\right) E_{\theta,\gamma,\nu}(T_{(n)})\right\} \\
& + \frac{1}{\lambda_2 n}\left\{\delta_1 E_{\theta,\gamma,\nu}(Z_1 T_{(1)}) + \delta_2 E_{\theta,\gamma,\nu}(Z_1 T_{(n)})\right\} + O\left(\frac{1}{n\sqrt{n}}\right).
\end{aligned}
\tag{3.27}
$$

Substituting (3.16), (3.17), (3.20) and (3.21) into (3.27), we obtain

$$
E_{\theta,\gamma,\nu}(\hat{U}) = -\frac{1}{\sqrt{\lambda_2 n}}\left\{\frac{1}{k}\left(\frac{\partial \lambda_1}{\partial \gamma}\right) - \frac{1}{\tilde{k}}\left(\frac{\partial \lambda_1}{\partial \nu}\right) + \frac{\lambda_3}{2\lambda_2}\right\} + O\left(\frac{1}{n\sqrt{n}}\right). \tag{3.28}
$$

From (3.25), we have

$$
\begin{aligned}
&E_{\theta,\gamma,\nu}(\hat{U}^2)\\
=&E_{\theta,\gamma,\nu}(Z_1^2) - \frac{1}{\sqrt{\lambda_2 n}}\left\{2\left(\frac{\partial\lambda_1}{\partial\gamma}\right)E_{\theta,\gamma,\nu}(Z_1T_{(1)}) + 2\left(\frac{\partial\lambda_1}{\partial\nu}\right)E_{\theta,\gamma,\nu}(Z_1T_{(n)})\right.\\
&\left.+\frac{\lambda_3}{\lambda_2}E_{\theta,\gamma,\nu}(Z_1^3)\right\} + \frac{1}{\lambda_2 n}\left\{\left(\frac{\partial\lambda_1}{\partial\gamma}\right)^2 E_{\theta,\gamma,\nu}(T_{(1)}^2)\right.\\
&\left.+2\left(\frac{\partial\lambda_1}{\partial\gamma}\right)\left(\frac{\partial\lambda_1}{\partial\nu}\right)E_{\theta,\gamma,\nu}(T_{(1)}T_{(n)}) + \left(\frac{\partial\lambda_1}{\partial\nu}\right)^2 E_{\theta,\gamma,\nu}(T_{(n)}^2)\right\}\\
&+\frac{1}{\lambda_2 n}\left\{\frac{\lambda_3}{\lambda_2}\left(\frac{\partial\lambda_1}{\partial\gamma}\right)+2\delta_1\right\}E_{\theta,\gamma,\nu}(Z_1^2T_{(1)}) + \frac{1}{\lambda_2 n}\left\{\frac{\lambda_3}{\lambda_2}\left(\frac{\partial\lambda_1}{\partial\nu}\right)+2\delta_2\right\}\\
&\cdot E_{\theta,\gamma,\nu}(Z_1^2T_{(n)}) + \frac{\lambda_3^2}{4\lambda_2^3 n}E_{\theta,\gamma,\nu}(Z_1^4) + \frac{1}{n}\left(\frac{\lambda_3^2}{\lambda_2^3} - \frac{\lambda_4}{3\lambda_2^2}\right)E_{\theta,\gamma,\nu}(Z_1^4)\\
&+O\left(\frac{1}{n\sqrt{n}}\right). \qquad (3.29)
\end{aligned}
$$

Since $T_{(1)}$ and $T_{(n)}$ are asymptotically independent, it follows from (3.16) and (3.17) that

$$
\begin{aligned}
E_{\theta,\gamma,\nu}(T_{(1)}T_{(n)}) &= E_{\theta,\gamma,\nu}(T_{(1)})E_{\theta,\gamma,\nu}(T_{(n)}) + O\left(\frac{1}{n}\right)\\
&= -\frac{1}{k\tilde{k}} + O\left(\frac{1}{n}\right). \qquad (3.30)
\end{aligned}
$$

Substituting (3.18)–(3.23), (3.26), and (3.30) into (3.29) we obtain

$$
\begin{aligned}
E_{\theta,\gamma,\nu}(\hat{U}^2) =&1 - \frac{2}{\lambda_2 n}\left[\frac{1}{k}\left(\frac{\partial\lambda_1}{\partial\gamma}\right)\left\{u(\gamma) - \lambda_1 + \frac{1}{k}\left(\frac{\partial\lambda_1}{\partial\gamma}\right)\right\}\right.\\
&\left.-\frac{1}{\tilde{k}}\left(\frac{\partial\lambda_1}{\partial\nu}\right)\left\{u(\nu) - \lambda_1 - \frac{1}{\tilde{k}}\left(\frac{\partial\lambda_1}{\partial\nu}\right)\right\}\right]\\
&-\frac{2}{k\tilde{k}\lambda_2 n}\left(\frac{\partial\lambda_1}{\partial\gamma}\right)\left(\frac{\partial\lambda_1}{\partial\nu}\right) + \frac{3\lambda_3}{\lambda_2^2 n}\left\{\frac{1}{k}\left(\frac{\partial\lambda_1}{\partial\gamma}\right) - \frac{1}{\tilde{k}}\left(\frac{\partial\lambda_1}{\partial\nu}\right)\right\}\\
&-\frac{2}{\lambda_2 n}\left\{\frac{1}{k}\left(\frac{\partial\lambda_2}{\partial\gamma}\right) - \frac{1}{\tilde{k}}\left(\frac{\partial\lambda_2}{\partial\nu}\right)\right\} + \frac{11\lambda_3^2}{4\lambda_2^3 n} - \frac{\lambda_4}{\lambda_2^2 n} + O\left(\frac{1}{n^2}\right).
\end{aligned}
\qquad (3.31)
$$

Next, put $\hat{\lambda}_2 = \lambda_2(\hat{\theta}_{ML}, X_{(1)}, X_{(n)})$, $\hat{k} = k(\hat{\theta}_{ML}, X_{(1)}, X_{(n)})$, and $\hat{\tilde{k}} = \tilde{k}(\hat{\theta}_{ML}, X_{(1)}, X_{(n)})$. Letting

$$
\frac{\partial\hat{\lambda}_1}{\partial\gamma} = \frac{\partial\lambda_1}{\partial\gamma}(\hat{\theta}_{ML}, X_{(1)}, X_{(n)}), \quad \frac{\partial\hat{\lambda}_1}{\partial\nu} = \frac{\partial\lambda_1}{\partial\nu}(\hat{\theta}_{ML}, X_{(1)}, X_{(n)}),
$$

we have

$$\frac{\sqrt{\lambda_2}}{\hat{\lambda}_2}\left(\frac{1}{\hat{k}}\frac{\partial\hat{\lambda}_1}{\partial\gamma}-\frac{1}{\hat{\tilde{k}}}\frac{\partial\hat{\lambda}_1}{\partial\nu}\right)$$
$$=\frac{1}{\sqrt{\lambda_2}}\left\{\frac{1}{k}\left(\frac{\partial\lambda_1}{\partial\gamma}\right)-\frac{1}{\tilde{k}}\left(\frac{\partial\lambda_1}{\partial\nu}\right)+\frac{1}{\sqrt{\lambda_2 n}}\left(\frac{1}{k}\left(\frac{\partial^2\lambda_1}{\partial\gamma\partial\theta}\right)-\frac{1}{\tilde{k}}\left(\frac{\partial^2\lambda_1}{\partial\nu\partial\theta}\right)\right)\hat{U}\right.$$
$$-\frac{1}{\sqrt{\lambda_2 n}}\left(\frac{1}{k^2}\left(\frac{\partial k}{\partial\theta}\right)\left(\frac{\partial\lambda_1}{\partial\gamma}\right)-\frac{1}{\tilde{k}^2}\left(\frac{\partial\tilde{k}}{\partial\theta}\right)\left(\frac{\partial\lambda_1}{\partial\nu}\right)\right)\hat{U}$$
$$\left.-\frac{\lambda_3}{\lambda_2^{3/2}\sqrt{n}}\left(\frac{1}{k}\left(\frac{\partial\lambda_1}{\partial\gamma}\right)-\frac{1}{\tilde{k}}\left(\frac{\partial\lambda_1}{\partial\nu}\right)\right)\hat{U}+O_p\left(\frac{1}{n}\right)\right\}. \tag{3.32}$$

From (3.5) and (3.32), we obtain the stochastic expansion

$$\hat{U}^* :=\sqrt{\lambda_2 n}(\hat{\theta}_{ML^*}-\theta)=\sqrt{\lambda_2 n}(\hat{\theta}_{ML}-\theta)+\frac{\sqrt{\lambda_2}}{\hat{\lambda}_2\sqrt{n}}\left\{\frac{1}{\hat{k}}\left(\frac{\partial\hat{\lambda}_1}{\partial\gamma}\right)-\frac{1}{\hat{\tilde{k}}}\left(\frac{\partial\hat{\lambda}_1}{\partial\nu}\right)\right\}$$
$$=\hat{U}+\frac{1}{\sqrt{\lambda_2 n}}\left\{\frac{1}{k}\left(\frac{\partial\lambda_1}{\partial\gamma}\right)-\frac{1}{\tilde{k}}\left(\frac{\partial\lambda_1}{\partial\nu}\right)\right\}$$
$$-\frac{1}{\lambda_2 n}\left\{\frac{\delta_1}{k}-\frac{\delta_2}{\tilde{k}}+\frac{1}{k^2}\left(\frac{\partial k}{\partial\theta}\right)\left(\frac{\partial\lambda_1}{\partial\gamma}\right)-\frac{1}{\tilde{k}^2}\left(\frac{\partial\tilde{k}}{\partial\theta}\right)\left(\frac{\partial\lambda_1}{\partial\nu}\right)\right\}Z_1$$
$$+O_p\left(\frac{1}{n\sqrt{n}}\right) \tag{3.33}$$

where $\hat{U}$ is given by (3.25). From (3.28), we have

$$E_{\theta,\gamma,\nu}(\hat{U}^*)=-\frac{\lambda_3}{2\lambda_2^{3/2}\sqrt{n}}+O\left(\frac{1}{n\sqrt{n}}\right). \tag{3.34}$$

Hence, we obtain (3.6) and (3.9). From (3.28), (3.30), and (3.33), we have

$$E_{\theta,\gamma,\nu}(\hat{U}^{*^2})=1-\frac{2}{k\lambda_2 n}\{u(\gamma)-\lambda_1\}\left(\frac{\partial\lambda_1}{\partial\gamma}\right)+\frac{2}{\tilde{k}\lambda_2 n}\{u(\nu)-\lambda_1\}\left(\frac{\partial\lambda_1}{\partial\nu}\right)$$
$$-\frac{3}{\lambda_2 n}\left\{\frac{1}{k^2}\left(\frac{\partial\lambda_1}{\partial\gamma}\right)^2+\frac{1}{\tilde{k}^2}\left(\frac{\partial\lambda_1}{\partial\nu}\right)^2\right\}$$
$$-\frac{2}{\lambda_2 n}\left\{\frac{1}{k^2}\left(\frac{\partial k}{\partial\theta}\right)\left(\frac{\partial\lambda_1}{\partial\gamma}\right)-\frac{1}{\tilde{k}^2}\left(\frac{\partial\tilde{k}}{\partial\theta}\right)\left(\frac{\partial\lambda_1}{\partial\nu}\right)\right\}$$
$$+\frac{11\lambda_3^2}{4\lambda_2^3 n}-\frac{\lambda_4}{\lambda_2^2 n}+O\left(\frac{1}{n\sqrt{n}}\right),$$

hence, by (3.34)

$$\begin{aligned}V_{\theta,\gamma,\nu}(\hat{U}^*) =&1+\frac{1}{n}\left(\frac{5\lambda_3^2}{2\lambda_2^3}-\frac{\lambda_4}{\lambda_2^2}\right)-\frac{2}{k\lambda_2 n}\left(\frac{\partial\lambda_1}{\partial\gamma}\right)\left\{u(\gamma)-\lambda_1+\frac{1}{k}\left(\frac{\partial k}{\partial\theta}\right)\right\}\\&-\frac{3}{k^2\lambda_2 n}\left(\frac{\partial\lambda_1}{\partial\gamma}\right)^2+\frac{2}{\tilde{k}\lambda_2 n}\left(\frac{\partial\lambda_1}{\partial\nu}\right)\left\{u(\gamma)-\lambda_1+\frac{1}{\tilde{k}}\left(\frac{\partial\tilde{k}}{\partial\theta}\right)\right\}\\&-\frac{3}{\tilde{k}^2\lambda_2 n}\left(\frac{\partial\lambda_1}{\partial\nu}\right)^2+O\left(\frac{1}{n^2}\right).\end{aligned} \tag{3.35}$$

Since

$$\lambda_1(\theta,\gamma,\nu)=\frac{\partial}{\partial\theta}\log b(\theta,\gamma,\nu)=\frac{1}{b(\theta,\gamma,\nu)}\int_\gamma^\nu u(x)a(x)e^{\theta u(x)}dx,$$

it follows that

$$\frac{\partial\lambda_1(\theta,\gamma,\nu)}{\partial\gamma}=\frac{a(\gamma)e^{\theta u(\gamma)}}{b(\theta,\gamma,\nu)}\{\lambda_1(\theta,\gamma,\nu)-u(\gamma)\}=k(\theta,\gamma,\nu)\{\lambda_1(\theta,\gamma,\nu)-u(\gamma)\}, \tag{3.36}$$

$$\frac{\partial\lambda_1(\theta,\gamma,\nu)}{\partial\nu}=\frac{a(\nu)e^{\theta u(\nu)}}{b(\theta,\gamma,\nu)}\{u(\nu)-\lambda_1(\theta,\gamma,\nu)\}=\tilde{k}(\theta,\gamma,\nu)\{u(\nu)-\lambda_1(\theta,\gamma,\nu)\}. \tag{3.37}$$

Since

$$\frac{\partial k}{\partial\theta}(\theta,\gamma,\nu)=k(\theta,\gamma,\nu)\{u(\gamma)-\lambda_1(\theta,\gamma,\nu)\}, \tag{3.38}$$

$$\frac{\partial\tilde{k}}{\partial\theta}(\theta,\gamma,\nu)=\tilde{k}(\theta,\gamma,\nu)\{u(\nu)-\lambda_1(\theta,\gamma,\nu)\}, \tag{3.39}$$

it is seen from (3.35)–(3.39) that

$$\begin{aligned}V_{\theta,\gamma,\nu}(\hat{U}^*) =&1+\frac{1}{n}\left(\frac{5\lambda_3^2}{2\lambda_2^3}-\frac{\lambda_4}{\lambda_2^2}\right)+\frac{1}{\lambda_2 n}\left[\{u(\gamma)-\lambda_1\}^2+\{u(\nu)-\lambda_1\}^2\right]\\&+O\left(\frac{1}{n\sqrt{n}}\right),\end{aligned}$$

hence we obtain (3.10). Thus, we complete the proof.

The proof of Theorem 3.5.1 Since, from (3.12)

$$
\begin{aligned}
0 =& \frac{1}{n-2}\sum_{i=2}^{n-1}\{u(Y_i) - \lambda_1(\theta, x_{(1)}, x_{(n)})\} - \frac{1}{\sqrt{n}}\lambda_2(\theta, x_{(1)}, x_{(n)})\sqrt{n}(\hat{\theta}_{MCL} - \theta) \\
& - \frac{1}{2n}\lambda_3(\theta, x_{(1)}, x_{(n)})n(\hat{\theta}_{MCL} - \theta)^2 - \frac{1}{6n\sqrt{n}}\lambda_4(\theta, x_{(1)}, x_{(n)})n\sqrt{n}(\hat{\theta}_{MCL} - \theta)^3 \\
& + O_p\left(\frac{1}{n^2}\right),
\end{aligned}
$$

letting

$$
\tilde{Z}_1 := \frac{1}{\sqrt{\tilde{\lambda}_2(n-2)}}\sum_{i=2}^{n-1}\{u(Y_i) - \lambda_1(\theta, x_{(1)}, x_{(n)})\}, \quad \tilde{U} := \sqrt{\tilde{\lambda}_2 n}(\hat{\theta}_{MCL} - \theta)
$$

where $\tilde{\lambda}_i = \lambda_i(\theta, x_{(1)}, x_{(n)})$ $(i = 1, 2, 3, 4)$, we have

$$
0 = \sqrt{\frac{\tilde{\lambda}_2}{n-2}}\tilde{Z}_1 - \sqrt{\frac{\tilde{\lambda}_2}{n}}\tilde{U} - \frac{\tilde{\lambda}_3}{2\tilde{\lambda}_2 n}\tilde{U}^2 - \frac{\tilde{\lambda}_4}{6\tilde{\lambda}_2^{3/2} n\sqrt{n}}\tilde{U}^3 + O_p\left(\frac{1}{n^2}\right),
$$

hence the stochastic expansion of $\hat{U}$ is given by

$$
\tilde{U} = \tilde{Z}_1 - \frac{\tilde{\lambda}_3}{2\tilde{\lambda}_2^{3/2}\sqrt{n}}\tilde{Z}_1^2 + \frac{1}{n}\tilde{Z}_1 + \frac{\tilde{\lambda}_3^2}{2\tilde{\lambda}_2^3 n}\tilde{Z}_1^3 - \frac{\tilde{\lambda}_4}{6\tilde{\lambda}_2^2 n}\tilde{Z}_1^3 + O_p\left(\frac{1}{n\sqrt{n}}\right). \tag{3.40}
$$

Since for $i = 2, 3, 4$

$$
\tilde{\lambda}_i = \lambda_i(\theta, X_{(1)}, X_{(n)}) = \lambda_i(\theta, \gamma, \nu) + \frac{1}{n}\left(\frac{\partial\lambda_i}{\partial\gamma}\right)T_{(1)} + \frac{1}{n}\left(\frac{\partial\lambda_i}{\partial\nu}\right)T_{(n)} + O_p\left(\frac{1}{n^2}\right),
$$

we obtain

$$
\begin{aligned}
\tilde{U} =& \sqrt{\tilde{\lambda}_2 n}(\hat{\theta}_{MCL} - \theta) \\
=& \sqrt{\lambda_2 n}(\hat{\theta}_{MCL} - \theta)\left\{1 + \frac{1}{2n\lambda_2}\left(\frac{\partial\lambda_2}{\partial\gamma}\right)T_{(1)} + \frac{1}{2n\lambda_2}\left(\frac{\partial\lambda_2}{\partial\nu}\right)T_{(n)} + O_p\left(\frac{1}{n^2}\right)\right\}
\end{aligned} \tag{3.41}
$$

where $T_{(1)} = n(X_{(1)} - \gamma)$, $T_{(n)} = n(X_{(n)} - \nu)$, and $\lambda_2 = \lambda_2(\theta, \gamma, \nu)$. Then, it follows from (3.40) and (3.41) that

$$\begin{aligned}\tilde{U}_0 &= \sqrt{\lambda_2 n}(\hat{\theta}_{MCL} - \theta)\\&= \tilde{Z}_1 - \frac{\tilde{\lambda}_3}{2\tilde{\lambda}_2^{3/2}\sqrt{n}}\tilde{Z}_1^2 + \frac{1}{n}\left\{1 - \frac{1}{2\lambda_2}\left(\frac{\partial \lambda_2}{\partial \gamma}\right)T_{(1)} - \frac{1}{2\lambda_2}\left(\frac{\partial \lambda_2}{\partial \nu}\right)T_{(n)}\right\}\tilde{Z}_1\\&\quad + \frac{1}{2n}\left(\frac{\tilde{\lambda}_3^2}{\tilde{\lambda}_2^3} - \frac{\tilde{\lambda}_4}{3\tilde{\lambda}_2^2}\right)\tilde{Z}_1^3 + O_p\left(\frac{1}{n\sqrt{n}}\right),\end{aligned} \tag{3.42}$$

hence we obtain (3.13). For given $X_{(1)} = x_{(1)}$ and $X_{(n)} = x_{(n)}$, i.e., $T_{(1)} = t_{(1)} := n(x_{(1)} - \gamma)$ and $T_{(n)} = t_{(n)} := n(x_{(n)} - \nu)$, the conditional expectations of $\tilde{Z}_1$ and $\tilde{Z}_1^2$ are

$$E_{\theta,\gamma,\nu}(\tilde{Z}_1 | t_{(1)}, t_{(n)}) = \frac{1}{\sqrt{\tilde{\lambda}_2(n-2)}}\sum_{i=2}^{n-1}\{E_{\theta,\gamma,\nu}[u(Y_i)|t_{(1)}, t_{(n)}] - \lambda_1(\theta, x_{(1)}, x_{(n)})\} = 0,$$

$$\begin{aligned}&E_{\theta,\gamma,\nu}(\tilde{Z}_1^2 | t_{(1)}, t_{(n)})\\&= \frac{1}{\tilde{\lambda}_2(n-2)}\Bigg[\sum_{i=2}^{n-1} E_{\theta,\gamma,\nu}[\{u(Y_i) - \lambda_1(\theta, x_{(1)}, x_{(n)})\}^2 | t_{(1)}, t_{(n)}]\\&\quad + \sum_{\substack{i\neq j\\2\le i,j\le n-1}}\sum E_{\theta,\gamma,\nu}\left[\{u(Y_i) - \lambda_1(\theta, x_{(1)}, x_{(n)})\}\{u(Y_j) - \lambda_1(\theta, x_{(1)}, x_{(n)})\} \,|\, t_{(1)}, t_{(n)}\right]\Bigg]\\&= 1,\end{aligned} \tag{3.43}$$

hence the conditional variance of $\tilde{Z}_1$ is equal to 1, i.e., $V_{\theta,\gamma,\nu}(\tilde{Z}_1 | t_{(1)}, t_{(n)}) = 1$. In a similar way to the above, we have

$$E_{\theta,\gamma,\nu}(\tilde{Z}_1^3 | t_{(1)}, t_{(n)}) = \frac{\tilde{\lambda}_3}{\tilde{\lambda}_2^{3/2}\sqrt{n-2}}, \quad E_{\theta,\gamma,\nu}(\tilde{Z}_1^4 | t_{(1)}, t_{(n)}) = 3 + \frac{\tilde{\lambda}_4}{\tilde{\lambda}_2^2(n-2)}. \tag{3.44}$$

Then, it follows from (3.43) and (3.44) that

$$E_{\theta,\gamma,\nu}(\tilde{U}_0 | T_{(1)}, T_{(n)}) = -\frac{\tilde{\lambda}_3}{2\tilde{\lambda}_2^{3/2}\sqrt{n}} + O_p\left(\frac{1}{n\sqrt{n}}\right), \tag{3.45}$$

$$\begin{aligned}E_{\theta,\gamma,\nu}(\tilde{U}_0^2 | T_{(1)}, T_{(n)}) &= 1 + \frac{2}{n} + \frac{11\tilde{\lambda}_3^2}{4\tilde{\lambda}_2^3 n} - \frac{\tilde{\lambda}_4}{\tilde{\lambda}_2^2 n} - \frac{1}{\lambda_2 n}\left(\frac{\partial \lambda_2}{\partial \gamma}\right)T_{(1)}\\&\quad - \frac{1}{\lambda_2 n}\left(\frac{\partial \lambda_2}{\partial \nu}\right)T_{(n)} + O_p\left(\frac{1}{n\sqrt{n}}\right)\end{aligned} \tag{3.46}$$

where $\tilde{\lambda}_i = \lambda_i(\theta, X_{(1)}, X_{(n)})$ $(i = 2, 3, 4)$. Since, for $i = 2, 3, 4$

$$\tilde{\lambda}_i = \lambda_i(\theta, X_{(1)}, X_{(n)}) = \lambda_i(\theta, \gamma, \nu) + O_p\left(\frac{1}{n}\right) = \lambda_i + O_p\left(\frac{1}{n}\right), \tag{3.47}$$

it follows from (3.45) that

$$E_{\theta,\gamma,\nu}(\tilde{U}_0) = E_{\theta,\gamma,\nu}[E_{\theta,\gamma,\nu}(\tilde{U}_0|T_{(1)}, T_{(n)})] = -\frac{\lambda_3}{2\lambda_2^{3/2}\sqrt{n}} + O\left(\frac{1}{n\sqrt{n}}\right). \tag{3.48}$$

In a similar way to the above, we obtain from (3.16), (3.17), (3.46) and (3.47) that

$$E_{\theta,\gamma,\nu}(\tilde{U}_0^2) = 1 + \frac{2}{n} + \frac{11\lambda_3^2}{4\lambda_2^3 n} - \frac{\lambda_4}{\lambda_2^2 n} - \frac{1}{k\lambda_2 n}\left(\frac{\partial\lambda_2}{\partial\gamma}\right) + \frac{1}{\tilde{k}\lambda_2 n}\left(\frac{\partial\lambda_2}{\partial\nu}\right) + O\left(\frac{1}{n\sqrt{n}}\right). \tag{3.49}$$

Since, by (3.36)–(3.39)

$$\frac{1}{k}\left(\frac{\partial\lambda_2}{\partial\gamma}\right) = \frac{1}{k}\left[\frac{\partial k}{\partial\theta}(\lambda_1 - u(\gamma)) + k\left(\frac{\partial\lambda_1}{\partial\theta}\right)\right] = -(u(\gamma) - \lambda_1)^2 + \lambda_2,$$
$$\frac{1}{\tilde{k}}\left(\frac{\partial\lambda_2}{\partial\nu}\right) = \frac{1}{\tilde{k}}\left[\frac{\partial\tilde{k}}{\partial\theta}(u(\nu) - \lambda_1) - \tilde{k}\left(\frac{\partial\lambda_1}{\partial\theta}\right)\right] = (u(\nu) - \lambda_1)^2 - \lambda_2,$$

it follows from (3.49) that

$$E_{\theta,\gamma,\nu}(\tilde{U}_0^2) = 1 + \frac{11\lambda_3^2}{4\lambda_2^3 n} - \frac{\lambda_4}{\lambda_2^2 n} + \frac{1}{\lambda_2 n}\left[\{u(\gamma) - \lambda_1\}^2 + \{u(\nu) - \lambda_1\}^2\right] + O\left(\frac{1}{n\sqrt{n}}\right),$$

hence, by (3.48)

$$V_{\theta,\gamma,\nu}(\tilde{U}_0) = 1 + \frac{1}{n}\left(\frac{5\lambda_3^2}{2\lambda_2^3} - \frac{\lambda_4}{\lambda_2^2}\right) + \frac{1}{\lambda_2 n}\left[\{u(\gamma) - \lambda_1\}^2 + \{u(\nu) - \lambda_1\}^2\right] + O\left(\frac{1}{n\sqrt{n}}\right).$$

Thus, we complete the proof.

3.10 Appendix B2

The proof of Lemma 3.9.1 The derivation of (3.14) and (3.15) is omitted, since it is essentially same as that of (2.15) in Lemma 2.9.1. The Eqs. (3.16)–(3.19) are obtained by straightforward calculation.

The proof of Lemma 3.9.2 Let $Y_1, \dots, Y_{n-1}$ be a random permutation of the $(n-1)!$ permutations of $X_{(1)}, \dots, X_{(n-1)}$ such that conditionally on $X_{(n)} = x_{(n)}$, $Y_1, \dots, Y_{n-1}$ are i.i.d. random variables according to a distribution with density

$$g(y; \theta, \gamma, x_{(n)}) = \frac{a(y)e^{\theta u(y)}}{b(\theta, \gamma, x_{(n)})} \quad \text{for } c < \gamma \le y < x_{(n)} \le \nu < d \tag{3.50}$$

with respect to the Lebesgue measure. Then, the conditional expectation of Z_1, given $T_{(n)}$, is obtained by

$$\begin{aligned} E_{\theta,\gamma,\nu}(Z_1|T_{(n)}) &= \frac{1}{\sqrt{\lambda_2 n}} \sum_{i=1}^{n} \{E_{\theta,\gamma,\nu}[u(X_i)|T_{(n)}] - \lambda_1\} \\ &= \frac{1}{\sqrt{\lambda_2 n}} \left\{ u(X_{(n)}) + \sum_{i=1}^{n-1} E_{\theta,\gamma,\nu}[u(Y_i)|T_{(n)}] - n\lambda_1 \right\} \end{aligned} \tag{3.51}$$

where $\lambda_i = \lambda_i(\theta, \gamma, \nu)$ $(i = 1, 2)$. Since, for $i = 1, \dots, n-1$,

$$E_{\theta,\gamma,\nu}[u(Y_i)|T_{(n)}] = \frac{\partial}{\partial\theta} \log b(\theta, \gamma, X_{(n)}) = \lambda_1(\theta, \gamma, X_{(n)}) =: \hat{\lambda}_1^{(n)} \quad \text{(say)},$$

it follows from (3.51) that

$$E_{0,\gamma,\nu}(Z_1|T_{(n)}) = \frac{1}{\sqrt{\lambda_2 n}} \left\{ u(X_{(n)}) + (n-1)\hat{\lambda}_1^{(n)} \right\} - \frac{\lambda_1\sqrt{n}}{\sqrt{\lambda_2}},$$

hence, from (3.17) and (3.51)

$$\begin{aligned} E_{\theta,\gamma,\nu}(Z_1 T_{(n)}) = &\frac{1}{\sqrt{\lambda_2 n}} \left\{ E_{\theta,\gamma,\nu}[u(X_{(n)})T_{(n)}] + (n-1)E_{\theta,\gamma,\nu}(\hat{\lambda}_1^{(n)} T_{(n)}) \right\} \\ &- \sqrt{\frac{n}{\lambda_2}}\lambda_1 \left\{ -\frac{1}{\tilde{k}} - \frac{\tilde{A}}{n} + O\left(\frac{1}{n^2}\right) \right\} \end{aligned} \tag{3.52}$$

where $\tilde{k} = \tilde{k}(\theta, \gamma, \nu)$ and $\tilde{A} = \tilde{A}(\theta, \gamma, \nu)$. Since, by the Taylor expansion

$$\begin{aligned} u(X_{(n)}) &= u(\nu) + \frac{u'(\nu)}{n} T_{(n)} + O_p\left(\frac{1}{n^2}\right), \\ \hat{\lambda}_1^{(n)} &= \lambda_1(\theta, \gamma, \nu) + \frac{1}{n}\left\{\frac{\partial}{\partial\nu}\lambda_1(\theta, \gamma, \nu)\right\} T_{(n)} + \frac{1}{2n^2}\left\{\frac{\partial^2}{\partial\nu^2}\lambda_1(\theta, \gamma, \nu)\right\} T_{(n)}^2 \\ &\quad + O_p\left(\frac{1}{n^3}\right), \end{aligned}$$

it follows from (3.17) and (3.19) that

$$E_{\theta,\gamma,\nu}[u(X_{(n)})T_{(n)}] = -\frac{u(\nu)}{\tilde{k}} - \left\{\tilde{A}u(\nu) - \frac{2u'(\nu)}{\tilde{k}^2}\right\}\frac{1}{n} + O\left(\frac{1}{n^2}\right), \tag{3.53}$$

$$E_{\theta,\gamma,\nu}(\hat{\lambda}_1^{(n)}T_{(n)}) = -\frac{\lambda_1}{\tilde{k}} - \left\{\lambda_1\tilde{A} - \frac{2}{\tilde{k}^2}\left(\frac{\partial\lambda_1}{\partial\nu}\right)\right\}\frac{1}{n} + O\left(\frac{1}{n^2}\right) \tag{3.54}$$

where $\tilde{k} = \tilde{k}(\theta, \gamma, \nu)$, $\tilde{A} = \tilde{A}(\theta, \gamma, \nu)$, and $\lambda_1 = \lambda_1(\theta, \gamma, \nu)$. From (3.52) – (3.54), we obtain

$$E_{\theta,\gamma,\nu}(Z_1T_{(n)}) = -\frac{1}{\tilde{k}\sqrt{\lambda_2 n}}\left\{u(\nu) - \lambda_1 - \frac{2}{\tilde{k}}\left(\frac{\partial\lambda_1}{\partial\nu}\right)\right\} + O\left(\frac{1}{n\sqrt{n}}\right).$$

On the other hand, it is shown in a similar way to Lemma 2.9.2 that

$$E_{\theta,\gamma,\nu}(Z_1T_{(1)}) = \frac{1}{k\sqrt{\lambda_2 n}}\left\{u(\gamma) - \lambda_1 + \frac{2}{k}\left(\frac{\partial\lambda_1}{\partial\gamma}\right)\right\} + O\left(\frac{1}{n\sqrt{n}}\right)$$

where $k = k(\theta, \gamma, \nu)$. Thus, we complete the proof.

The proof of Lemma 3.9.3 First, we have

$$\begin{aligned} E_{\theta,\gamma,\nu}(Z_1^2|T_{(n)}) =& \frac{1}{\lambda_2 n}\left\{u(X_{(n)}) - \lambda_1\right\}^2 \\ &+ \frac{2}{\lambda_2 n}\left\{u(X_{(n)}) - \lambda_1\right\}\sum_{i=1}^{n-1} E_{\theta,\gamma,\nu}\left[u(Y_i) - \lambda_1|T_{(n)}\right] \\ &+ \frac{1}{\lambda_2 n}\sum_{i=1}^{n-1} E_{\theta,\gamma,\nu}\left[\{u(Y_i) - \lambda_1\}^2\,|T_{(n)}\right] \\ &+ \frac{1}{\lambda_2 n}\sum_{\substack{i\neq j \\ 1\leq i,j\leq n-1}}\sum E_{\theta,\gamma,\nu}\left[\{u(Y_i) - \lambda_1\}\{u(Y_j) - \lambda_1\}\,|\,T_{(n)}\right]. \end{aligned} \tag{3.55}$$

For $1 \leq i \leq n-1$, we have

$$\begin{aligned} E_{\theta,\gamma,\nu}[u(Y_i) - \lambda_1|T_{(n)}] &= E_{\theta,\gamma,\nu}[u(Y_i)|T_{(n)}] - \lambda_1 = \lambda_1(\theta, \gamma, X_{(n)}) - \lambda_1(\theta, \gamma, \nu) \\ &= \hat{\lambda}_1^{(n)} - \lambda_1 = \left(\frac{\partial\lambda_1}{\partial\nu}\right)\frac{T_{(n)}}{n} + O_p\left(\frac{1}{n^2}\right) \end{aligned} \tag{3.56}$$

and, for $i \neq j$ and $1 \leq i, j \leq n-1$

$$E_{\theta,\gamma,\nu}\left[\{u(Y_i)-\lambda_1\}\{u(Y_j)-\lambda_1\}|T_{(n)}\right] = \left(\frac{\partial\lambda_1}{\partial\nu}\right)^2 \frac{T_{(n)}^2}{n^2} + O_p\left(\frac{1}{n^3}\right). \tag{3.57}$$

Since, for $i = 1, \ldots, n-1$

$$\begin{aligned} E_{\theta,\gamma,\nu}[u^2(Y_i)|T_{(n)}] &= \lambda_1^2(\theta,\gamma,X_{(n)}) + \lambda_2(\theta,\gamma,X_{(n)}) \\ &=: \hat{\lambda}_1^{(n)^2} + \hat{\lambda}_2^{(n)} \end{aligned}$$

where $\hat{\lambda}_i^{(n)} = \lambda_i(\theta,\gamma,X_{(n)})$ $(i = 1, 2)$, we have for $i = 1, \ldots, n-1$

$$E_{\theta,\gamma,\nu}[\{u(Y_i)-\lambda_1\}^2|T_{(n)}] = \lambda_2 + \frac{1}{n}\left(\frac{\partial\lambda_2}{\partial\nu}\right)T_{(n)} + O_p\left(\frac{1}{n^2}\right). \tag{3.58}$$

From (3.55)–(3.58), we obtain

$$E_{\theta,\gamma,\nu}(Z_1^2|T_{(n)}) = 1 + O_p\left(\frac{1}{n}\right), \tag{3.59}$$

hence, by (3.17)

$$E_{\theta,\gamma,\nu}(Z_1^2 T_{(n)}) = E_{\theta,\gamma,\nu}\left[T_{(n)}E_{\theta,\gamma,\nu}(Z_1^2|T_{(n)})\right] = -\frac{1}{\tilde{k}} + O\left(\frac{1}{n}\right).$$

On the other hand, it is shown a similar way to Lemma 2.9.3 that $E_{\theta,\gamma,\nu}(Z_1^2 T_{(1)}) = (1/k) + O(1/n)$. Thus, we complete the proof.

References

Aban, I. B., Meerschaert, M. M., & Panorska, A. K. (2006). Parameter estimation for the truncated Pareto distribution. *Journal of American Statistical Association*, *101*, 270–277.

Akahira, M., Hashimoto, S., Koike, K., & Ohyauchi, N. (2016). Second order asymptotic comparison of the MLE and MCLE for a two-sided truncated exponential family of distributions. *Communications in Statistics: Theory and Methods*, *45*, 5637–5659.

Beg, M. A. (1981). Estimation of the tail probability of the truncated Pareto distribution. *Journal of Information Optimization Sciences*, *2*, 192–198.

Hill, B. (1975). A simple general approach to inference about the tail of a distribution. *Annals of Statistics*, *3*, 1163–1173.

Chapter 4
Estimation of a Truncation Parameter for a One-Sided TEF

For a one-sided truncated exponential family (oTEF) of distributions with a truncation parameter γ and a natural parameter θ as a nuisance parameter, the maximum likelihood estimation on γ is discussed together with a bias-adjustment.

4.1 Introduction

In Chap. 2, for a oTEF with a natural parameter θ and a truncation parameter γ which is regarded as a typical non-regular case, we discussed a problem of estimating θ in the presence of γ as a nuisance parameter. In this chapter, following mostly the paper by Akahira and Ohyauchi (2017), we consider a problem of estimating γ in the presence of θ as a nuisance parameter in exchanging an interest parameter for a nuisance parameter. Let $\hat{\gamma}_{ML}^{\theta}$ and $\hat{\gamma}_{ML}$ be the MLEs of γ based on a sample of size n when θ is known and when θ is unknown, respectively. The stochastic expansions of the bias-adjusted MLEs $\hat{\gamma}_{ML^*}^{\theta}$ and $\hat{\gamma}_{ML^*}$ are derived, and the second-order asymptotic loss of $\hat{\gamma}_{ML^*}$ relative to $\hat{\gamma}_{ML^*}^{\theta}$ is also obtained. Further the truncated exponential and truncated normal, Pareto, lower-truncated beta, and lower-truncated Erlang type cases are discussed including consideration from the viewpoint of minimum variance unbiased estimation.

4.2 Preliminaries

According to Chap. 2, we have the formulation as follows. Suppose that $X_1, X_2, \cdots, X_n, \cdots$ is a sequence of i.i.d. random variables according to $P_{\theta,\gamma}$, having the density (1.7), which belongs to a oTEF $\mathscr{P}_o$. Then, we consider the estimation problem on γ in the presence of θ as a nuisance parameter. For any $\gamma \in (c, d)$, $\log b(\theta, \gamma)$ is strictly convex and infinitely differentiable in $\theta \in \Theta$ and

M. Akahira, *Statistical Estimation for Truncated Exponential Families*,
JSS Research Series in Statistics, DOI 10.1007/978-981-10-5296-5_4

$$\lambda_j(\theta, \gamma) := \frac{\partial^j}{\partial \theta^j} \log b(\theta, \gamma) \tag{4.1}$$

is the jth cumulant corresponding to (1.7) for $j = 1, 2, \cdots$. Let

$$k(\theta, \gamma) = a(\gamma)e^{\theta u(\gamma)}/b(\theta, \gamma), \tag{4.2}$$

$$A(\theta, \gamma) = -\frac{1}{k^2(\theta, \gamma)} \left\{ \frac{\partial}{\partial \gamma} \log k(\theta, \gamma) \right\} \tag{4.3}$$

which are defined in Theorem 2.4.1 and Lemma 2.9.1, respectively.

In the subsequent sections, we obtain the bias-adjusted MLE $\hat{\gamma}^{\theta}_{ML^*}$ and $\hat{\gamma}_{ML^*}$ of γ for known and unknown θ, respectively. Calculating their asymptotic variances based on their stochastic expansions, we get the second-order asymptotic loss of $\hat{\gamma}_{ML^*}$ relative to $\hat{\gamma}^{\theta}_{ML^*}$. Several examples are also given, and further, the proofs of theorems are located in Appendix C.

4.3 Bias-Adjusted MLE $\hat{\gamma}^{\theta}_{ML^*}$ of γ When θ is Known

For given $\boldsymbol{x}:=(x_1, \cdots, x_n)$ satisfying $\gamma \leq x_{(1)} := \min_{1 \leq i \leq n} x_i$ and $x_{(n)} := \max_{1 \leq i \leq n} x_i < d$, the likelihood function of γ is given by

$$L^{\theta}(\gamma; \boldsymbol{x}) = \frac{1}{b^n(\theta, \gamma)} \left\{ \prod_{i=1}^{n} a(x_i) \right\} \exp \left\{ \theta \sum_{i=1}^{n} u(x_i) \right\} \tag{4.4}$$

when θ is known. From (4.4), it follows that the MLE $\hat{\gamma}^{\theta}_{ML}$ of γ is given by $X_{(1)} := \min_{1 \leq i \leq n} X_i$. Let $T_{(1)} := n(X_{(1)} - \gamma)$. Then, we have the following.

Theorem 4.3.1 *For a oTEF $\mathscr{P}_o$ of distributions having densities of the form (1.7) with a truncation parameter γ and a natural parameter θ, let $\hat{\gamma}^{\theta}_{ML^*} = X^*_{(1)}$ be a bias-adjusted MLE of γ such that*

$$X^*_{(1)} := X_{(1)} - \frac{1}{\hat{k}_{\theta} n}, \tag{4.5}$$

where $\hat{k}_{\theta} = k(\theta, X_{(1)})$. Then, the stochastic expansion of $T^*_{(1)} := n(X^*_{(1)} - \gamma)$ is given by

$$T^*_{(1)} = T_{(1)} - \frac{1}{k} + \frac{1}{kn} \left(\frac{\partial}{\partial \gamma} \log k \right) T_{(1)} + O_p\left(\frac{1}{n^2} \right), \tag{4.6}$$

where $k = k(\theta, \gamma)$, and the second-order asymptotic mean and variance are given by

$$E_\gamma\left[T_{(1)}^*\right] = O\left(\frac{1}{n^2}\right), \tag{4.7}$$

$$V_\gamma\left(kT_{(1)}^*\right) = 1 - \frac{2}{kn}\left(\frac{\partial}{\partial\gamma}\log k\right) + O\left(\frac{1}{n^2}\right), \tag{4.8}$$

respectively.

4.4 Bias-Adjusted MLE $\hat{\gamma}_{ML^*}$ of γ When θ is Unknown

For given $\boldsymbol{x}$ satisfying $\gamma \le x_{(1)}$ and $x_{(n)} < d$, the likelihood function of γ and θ is given by

$$L(\gamma, \theta; \boldsymbol{x}) = \frac{1}{b^n(\theta, \gamma)}\left\{\prod_{i=1}^n a(x_i)\right\}\exp\left\{\theta\sum_{i=1}^n u(x_i)\right\}. \tag{4.9}$$

Let $\hat{\gamma}_{ML}$ and $\hat{\theta}_{ML}$ be the MLEs of γ and θ, respectively. From (4.9), it is seen that $\hat{\gamma}_{ML} = X_{(1)}$ and $L(X_{(1)}, \hat{\theta}_{ML}; \boldsymbol{X}) = \sup_{\theta\in\Theta} L(X_{(1)}, \theta; \boldsymbol{X})$, hence $\hat{\theta}_{ML}$ satisfies the likelihood equation

$$\frac{1}{n}\sum_{i=1}^n u(X_i) - \lambda_1(\hat{\theta}_{ML}, X_{(1)}) = 0, \tag{4.10}$$

where $\boldsymbol{X} = (X_1, \cdots, X_n)$. Let $\lambda_2 = \lambda_2(\theta, \gamma)$ and $\hat{U} = \sqrt{\lambda_2 n}(\hat{\theta}_{ML} - \theta)$. Then, we have the following.

Theorem 4.4.1 *For a oTEF $\mathscr{P}_o$ of distributions having densities of the form (1.7) with a truncation parameter γ and a natural parameter θ, let $\hat{\gamma}_{ML^*} = X_{(1)}^{**}$ be a bias-adjusted MLE of γ such that*

$$\begin{aligned} X_{(1)}^{**} := X_{(1)} &- \frac{1}{\hat{k}n} + \frac{1}{\hat{k}^2\hat{\lambda}_2 n^2}\left(\frac{\partial\hat{k}}{\partial\theta}\right)\left\{\frac{1}{\hat{k}}\left(\frac{\partial\hat{\lambda}_1}{\partial\gamma}\right) + \frac{\hat{\lambda}_3}{2\hat{\lambda}_2}\right\} \\ &- \frac{1}{2\hat{k}^2\hat{\lambda}_2 n^2}\left\{\frac{\partial^2\hat{k}}{\partial\theta^2} - \frac{2}{\hat{k}}\left(\frac{\partial\hat{k}}{\partial\theta}\right)^2\right\}, \end{aligned} \tag{4.11}$$

where $\hat{k} = k(\hat{\theta}_{ML}, X_{(1)})$, $\partial^j\hat{k}/\partial\theta^j = (\partial^j/\partial\theta^j)k(\hat{\theta}_{ML}, X_{(1)})$ $(j = 1, 2)$, $\hat{\lambda}_j = \lambda_j(\hat{\theta}_{ML}, X_{(1)})$ $(j = 2, 3)$ and $\partial\hat{\lambda}_1/\partial\gamma = (\partial/\partial\gamma)\lambda_1(\hat{\theta}_{ML}, X_{(1)})$. Then, the stochastic expansion of $T_{(1)}^{**} := n(X_{(1)}^{**} - \gamma)$ is given by

$$T_{(1)}^{**} = T_{(1)} - \frac{1}{k} + \frac{1}{k^2\sqrt{\lambda_2 n}}\left(\frac{\partial k}{\partial \theta}\right)\left\{\hat{U} + \frac{1}{\sqrt{\lambda_2 n}}\left(\frac{1}{k}\left(\frac{\partial \lambda_1}{\partial \gamma}\right) + \frac{\lambda_3}{2\lambda_2}\right)\right\}$$
$$+ \frac{1}{kn}\left(\frac{\partial}{\partial \gamma}\log k\right)T_{(1)} + \frac{1}{2k^2\lambda_2 n}\left[\frac{\partial^2 k}{\partial \theta^2} - \frac{2}{k}\left(\frac{\partial k}{\partial \theta}\right)^2\right]\left(\hat{U}^2 - 1\right)$$
$$+ O_p\left(\frac{1}{n\sqrt{n}}\right), \quad (4.12)$$

where $k = k(\theta, \gamma)$, $\lambda_j = \lambda_j(\theta, \gamma)$ $(j = 1, 2, 3)$, *and the second-order asymptotic mean and variance are given by*

$$E_{\theta,\gamma}\left[T_{(1)}^{**}\right] = O\left(\frac{1}{n\sqrt{n}}\right), \quad (4.13)$$
$$V_{\theta,\gamma}\left(kT_{(1)}^{**}\right) = 1 - \frac{2}{kn}\left(\frac{\partial}{\partial \gamma}\log k\right) + \frac{1}{\lambda_2 n}(u(\gamma) - \lambda_1)^2 + O\left(\frac{1}{n\sqrt{n}}\right). \quad (4.14)$$

4.5 Second-Order Asymptotic Loss of $\hat{\gamma}_{ML^*}$ Relative to $\hat{\gamma}_{ML^*}^{\theta}$

From the results in previous sections, we can asymptotically compare the bias-adjusted MLEs $\hat{\gamma}_{ML^*}^{\theta}$ and $\hat{\gamma}_{ML^*}$ of γ using their second-order asymptotic variances as follows.

Theorem 4.5.1 *For a oTEF* $\mathscr{P}_o$ *of distributions having densities of the form (1.7) with a truncation parameter* γ *and a natural parameter* θ*, let* $\hat{\gamma}_{ML^*}^{\theta}$ *and* $\hat{\gamma}_{ML^*}$ *be the bias-adjusted MLEs of* γ *when* θ *is known and when* θ *is unknown, respectively. Then, the second-order asymptotic loss of* $\hat{\gamma}_{ML^*} = X_{(1)}^{**}$ *relative to* $\hat{\gamma}_{ML^*}^{\theta} = X_{(1)}^{*}$ *is given by*

$$d_n\left(\hat{\gamma}_{ML^*}, \hat{\gamma}_{ML^*}^{\theta}\right) := n\left\{V_{\theta,\gamma}\left(kT_{(1)}^{**}\right) - V_{\gamma}\left(kT_{(1)}^{*}\right)\right\} = \frac{\{u(\gamma) - \lambda_1\}^2}{\lambda_2} + o(1) \quad (4.15)$$

as $n \to \infty$.

The proof is straightforward from Theorems 4.3.1 and 4.4.1.

Remark 4.5.1 It is seen from (1.6) and (4.15) that the ratio of the asymptotic variance of $kT_{(1)}^{**}$ to that of $kT_{(1)}^{*}$ is given by

$$R_n(\hat{\gamma}_{ML^*}, \hat{\gamma}_{ML^*}^{\theta}) = 1 + \frac{\{u(\gamma) - \lambda_1\}^2}{\lambda_2 n} + o\left(\frac{1}{n}\right).$$

From the consideration of models in Sect. 1.1, using (1.5), (1.6) and (4.15) we see that the difference between the asymptotic models $M(\hat{\gamma}_{ML^*}, \theta)$ and $M(\hat{\gamma}^{\theta}_{ML^*}, \theta)$ is given by $d_n(\hat{\gamma}_{ML^*}, \hat{\gamma}^{\theta}_{ML^*})$ or $R_n(\hat{\gamma}_{ML^*}, \hat{\gamma}^{\theta}_{ML^*})$ up to the second order, through the MLE of γ.

Remark 4.5.2 The second-order asymptotic loss (4.15) of $\hat{\gamma}_{ML^*}$ relative to $\hat{\gamma}^{\theta}_{ML^*}$ coincides with (2.8) of the bias-adjusted MLE $\hat{\theta}_{ML^*}$ of θ when γ is unknown relative to the MLE $\hat{\theta}^{\gamma}_{ML}$ of θ when γ is known, which seems to show a dual relation on the second-order asymptotic loss. It is noted that the standardization is necessary in the comparison.

Remark 4.5.3 Suppose that $X_1, X_2, \cdots, X_n, \cdots$ is a sequence of i.i.d. random variables according to an upper-truncated exponential family $\mathscr{P}'_o$ of distributions with densities of the form

$$f(x;\theta,\nu) = \begin{cases} \dfrac{a(x)e^{\theta u(x)}}{b(\theta,\nu)} & \text{for } c < x \le \nu < d, \\ 0 & \text{otherwise} \end{cases}$$

with respect to the Lebesgue measure, where $b(\theta, \nu)$ is a normalizing factor. Letting $Y_i = -X_i$ $(i = 1, 2, \cdots)$, and returning to the case of the lower-truncated exponential family with (1.7), we may obtain similar results to the above in a problem of estimating an upper truncation parameter ν in the presence of θ as a nuisance parameter.

4.6 Examples

Examples on the second-order asymptotic loss of the estimators are given for a lower-truncated exponential and a lower-truncated normal, Pareto, a lower-truncated beta, and a lower-truncated Erlang distributions, which are treated in Chap. 2.

Example 4.6.1 **(Lower-truncated exponential distribution)** (Continued from Example 2.7.1). Let $c = -\infty$, $d = \infty$, $a(x) \equiv 1$ and $u(x) = -x$ for $-\infty < \gamma \le x < \infty$ in the density (1.7). Since $b(\theta, \gamma) = e^{-\theta\gamma}/\theta$ for $\theta \in \Theta = (0, \infty)$, it follows from (4.1) that $\lambda_1(\theta, \gamma) = -\gamma - (1/\theta)$, $\lambda_2(\theta, \gamma) = 1/\theta^2$, $\lambda_3(\theta, \gamma) = -2/\theta^3$. Since, by (4.2), $k(\theta, \gamma) = \theta$, it is seen that $(\partial/\partial\theta)k(\theta, \gamma) = 1$, $(\partial^2/\partial\theta^2)k(\theta, \gamma) = 0$. When θ is known, it follows from (4.5) that the bias-adjusted MLE $\hat{\gamma}^{\theta}_{ML^*}$ of γ is given by $X^*_{(1)} = X_{(1)} - (\theta n)^{-1}$. When θ is unknown, it is seen from (4.10) that the MLE $\hat{\theta}_{ML}$ of θ is given by $\hat{\theta}_{ML} = 1/(\bar{X} - X_{(1)})$, hence by (4.11) the bias-adjusted MLE $\hat{\gamma}_{ML^*}$ of γ is given by

$$X^{**}_{(1)} = X_{(1)} - \left(\frac{1}{n} + \frac{1}{n^2}\right)(\bar{X} - X_{(1)}).$$

From Theorem 4.5.1, it follows that the second-order asymptotic loss of $\hat{\gamma}_{ML^*} = X_{(1)}^{**}$ for unknown θ relative to $\hat{\gamma}_{ML^*}^{\theta} = X_{(1)}^{*}$ for known θ is given by $d_n\left(\hat{\gamma}_{ML^*},\ \hat{\gamma}_{ML^*}^{\theta}\right) = 1 + o(1)$ as $n \to \infty$. Note that the loss is independent of θ up to the order $o(1)$. A similar consideration on the ratio $R_n(\hat{\gamma}_{ML^*}, \hat{\gamma}_{ML^*}^{\theta})$ to Example 2.7.1 is done.

In this case, we have the UMVU estimator

$$\hat{\gamma}_{UMVU} = X_{(1)} - \frac{1}{\theta n} = X_{(1)}^{*} = \hat{\gamma}_{ML^*}^{\theta}$$

when θ is known (see Voinov and Nikulin (1993)). When θ is unknown, we obtain

$$\hat{\gamma}_{UMVU} = X_{(1)} - \frac{1}{n-1}(\bar{X} - X_{(1)}),$$

hence

$$\hat{\gamma}_{ML^*} = X_{(1)}^{**} = \hat{\gamma}_{UMVU} + \frac{1}{n^2(n-1)}(\bar{X} - X_{(1)}).$$

Then $E_{\theta,\gamma}(\hat{\gamma}_{ML^*}) = \gamma + \theta^{-1}n^{-3}$ for any fixed n, hence $\hat{\gamma}_{ML^*}$ is not unbiased for γ. Since the variances of $\hat{\gamma}_{UMVU}^{\theta}$ and $\hat{\gamma}_{UMVU}$ are given by

$$V_{\gamma}\left(\hat{\gamma}_{UMVU}^{\theta}\right) = \frac{1}{\theta^2 n^2}, \quad \text{i.e.} \quad V_{\gamma}\left(\theta n \hat{\gamma}_{UMVU}^{\theta}\right) = 1$$

and

$$V_{\theta,\gamma}\left(\hat{\gamma}_{UMVU}\right) = \frac{1}{\theta^2 n(n-1)}, \quad \text{i.e.} \quad V_{\theta,\gamma}\left(\theta n \hat{\gamma}_{UMVU}\right) = 1 + \frac{1}{n-1},$$

we have the second-order asymptotic loss of $\hat{\gamma}_{UMVU}$ relative to $\hat{\gamma}_{UMVU}^{\theta}$

$$\begin{aligned} d_n\left(\hat{\gamma}_{UMVU}, \hat{\gamma}_{UMVU}^{\theta}\right) &= n\left\{V_{\theta,\gamma}\left(\theta n(\hat{\gamma}_{UMVU} - \gamma)\right) - V_{\gamma}\left(\theta n(\hat{\gamma}_{UMVU}^{\theta} - \gamma)\right)\right\} \\ &= 1 + o(1) \end{aligned}$$

as $n \to \infty$.

Example 4.6.2 **(Lower-truncated normal distribution)** (Continued from Example 2.7.2). Let $c = -\infty$, $d = \infty$, $a(x) = e^{-x^2/2}$ and $u(x) = x$ for $-\infty < \gamma \le x < \infty$ in the density (1.7). Since $b(\theta, \gamma) = \Phi(\theta - \gamma)/\phi(\theta)$ for $\theta \in \Theta = (-\infty, \infty)$, it follows from (4.1) that

$$\begin{aligned} &\lambda_1(\theta, \gamma) = \theta + \rho(\theta - \gamma), \qquad \frac{\partial \lambda_1(\theta, \gamma)}{\partial \gamma} = (\theta - \gamma)\rho(\theta - \gamma) + \rho^2(\theta - \gamma), \\ &\lambda_2(\theta, \gamma) = 1 - (\theta - \gamma)\rho(\theta - \gamma) - \rho^2(\theta - \gamma), \\ &\lambda_3(\theta, \gamma) = \rho(\theta - \gamma)\left\{2\rho^2(\theta - \gamma) + 3(\theta - \gamma)\rho(\theta - \gamma) + (\theta - \gamma)^2 - 1\right\}, \end{aligned}$$

where $\rho(t) := \phi(t)/\Phi(t)$ with $\Phi(x) = \int_{-\infty}^{x} \phi(t)dt$ and $\phi(t) = (1/\sqrt{2\pi})e^{-t^2/2}$ for $-\infty < t < \infty$. We also have from (4.2)

$$k(\theta, \gamma) = \rho(\theta - \gamma), \quad \frac{\partial k(\theta, \gamma)}{\partial \theta} = -(\theta - \gamma)\rho(\theta - \gamma) - \rho^2(\theta - \gamma),$$

$$\frac{\partial^2 k(\theta, \gamma)}{\partial \theta^2} = \rho(\theta - \gamma)\left\{2\rho^2(\theta - \gamma) + 3(\theta - \gamma)\rho(\theta - \gamma) + (\theta - \gamma)^2 - 1\right\},$$

$$\frac{\partial k(\theta, \gamma)}{\partial \gamma} = (\theta - \gamma)\rho(\theta - \gamma) + \rho^2(\theta - \gamma).$$

When θ is known, it follows from (4.5) that the bias-adjusted MLE $\hat{\gamma}^{\theta}_{ML^*}$ of γ is

$$X^*_{(1)} = X_{(1)} - \frac{1}{\rho(\theta - X_{(1)})n}.$$

When θ is unknown, it is seen from (4.10) that the MLE $\hat{\theta}_{ML}$ of θ satisfies the equation $\rho(\hat{\theta}_{ML} - X_{(1)}) = \bar{X} - \hat{\theta}_{ML}$, hence the bias-adjusted MLE $\hat{\gamma}_{ML^*}$ of γ is

$$X^{**}_{(1)} = X_{(1)} - \frac{1}{n(\bar{X} - \hat{\theta}_{ML})} + \frac{1 - (\bar{X} - X_{(1)})(\bar{X} - X_{(1)} + \bar{X} - \hat{\theta}_{ML})}{2n^2(\bar{X} - \hat{\theta}_{ML})\{1 - (\bar{X} - \hat{\theta}_{ML})(\bar{X} - X_{(1)})\}}.$$

From Theorem 4.5.1 it follows that the second-order asymptotic loss of $\hat{\gamma}_{ML^*}(= X^{**}_{(1)})$ for unknown θ relative to $\hat{\gamma}^{\theta}_{ML^*}(= X^*_{(1)})$ for known θ is given by

$$d_n\left(\hat{\gamma}_{ML^*}, \hat{\gamma}^{\theta}_{ML^*}\right) = \frac{\{\theta - \gamma + \rho(\theta - \gamma)\}^2}{1 - (\theta - \gamma)\rho(\theta - \gamma) - \rho^2(\theta - \gamma)} + o(1)$$

as $n \to \infty$. A similar consideration on the ratio $R_n(\hat{\gamma}_{ML^*}, \hat{\gamma}^{\theta}_{ML^*})$ to Example 2.7.2 in Sect. 2.7 is done.

Example 4.6.3 **(Pareto distribution)** (Continued from Example 2.7.3). Let $c = 0$, $d = \infty$, $a(x) = 1/x$ and $u(x) = -\log x$ for $0 < \gamma \le x < \infty$ in the density (1.7). Then $b(\theta, \gamma) = 1/(\theta\gamma^{\theta})$ for $\theta \in \Theta = (0, \infty)$, and it follows from (4.1) that $k(\theta, \gamma) = \theta/\gamma$, $\partial k/\partial\theta = 1/\gamma$ and $\partial k/\partial\gamma = -\theta/\gamma^2$. When θ is known, it follows from (4.5) that the bias-adjusted MLE $\hat{\gamma}^{\theta}_{ML^*}$ of γ is given by $X^*_{(1)} = \{1 - (\theta n)^{-1}\}X_{(1)}$, hence by (4.7) and (4.8)

$$E_{\gamma}\left[T^*_{(1)}\right] = O\left(\frac{1}{n^2}\right), \quad V_{\gamma}\left(\frac{\theta}{\gamma}T^*_{(1)}\right) = 1 + \frac{2}{\theta n} + O\left(\frac{1}{n^2}\right) \tag{4.16}$$

as $n \to \infty$, where $T^*_{(1)} = n(X^*_{(1)} - \gamma)$. On the other hand, in the Pareto case, it is known that the UMVU estimator of γ is given by $\hat{\gamma}^{\theta}_{UMVU} = X^*_{(1)}$ and its variance is $V_{\gamma}\left(\hat{\gamma}^{\theta}_{UMVU}\right) = \gamma^2/\{\theta n(\theta n - 2)\}$ (see, e.g., Voinov and Nikulin (1993)), hence

$$V_\gamma\left(\frac{\theta n}{\gamma}\hat{\gamma}_{UMVU}^{\theta}\right)=\frac{\theta n}{\theta n-2}=1+\frac{2}{\theta n}+O\left(\frac{1}{n^2}\right), \tag{4.17}$$

which is equal to (4.16) up to the order $1/n$ as $n\to\infty$.

Next, we consider the case when θ is unknown. Since $\partial^2 k/\partial\theta^2=0$, $\lambda_1=-(1/\theta)-\log\gamma$, $\lambda_2=1/\theta^2$, $\lambda_3=-2/\theta^3$ and $\partial\lambda_1/\partial\gamma=-1/\gamma$, it follows from (4.11) that the bias-adjusted MLE $\hat{\gamma}_{ML^*}$ of γ is given by

$$X_{(1)}^{**}=\left\{1-\left(\frac{1}{n}+\frac{1}{n^2}\right)\frac{1}{\hat{\theta}_{ML}}\right\}X_{(1)},$$

where $\hat{\theta}_{ML}=n/\sum_{i=1}^{n}\log(X_{(i)}/X_{(1)})$ from (4.10). Since $(\partial/\partial\gamma)\log k=-1/\gamma$, we have from (4.13) and (4.14)

$$E_{\theta,\gamma}\left[T_{(1)}^{**}\right]=O\left(\frac{1}{n\sqrt{n}}\right),\quad V_{\theta,\gamma}\left(\frac{\theta}{\gamma}T_{(1)}^{**}\right)=1+\frac{1}{n}\left(1+\frac{2}{\theta}\right)+O\left(\frac{1}{n^2}\right) \tag{4.18}$$

as $n\to\infty$, where $T_{(1)}^{**}=n(X_{(1)}^{**}-\gamma)$. On the other hand, in the Pareto case, it is known that the UMVU estimator of γ is given by

$$\hat{\gamma}_{UMVU}=X_{(1)}-\frac{X_{(1)}}{(n-1)\hat{\theta}_{ML}}$$

and its variance is

$$V_{\theta,\gamma}\left(\hat{\gamma}_{UMVU}\right)=\frac{\gamma^2}{\theta(n-1)(\theta n-2)}$$

(see, e.g., Voinov and Nikulin (1993)), hence

$$V_{\theta,\gamma}\left(\frac{\theta n}{\gamma}\hat{\gamma}_{UMVU}\right)=\frac{\theta n^2}{(n-1)(\theta n-2)}=1+\frac{1}{n}\left(1+\frac{2}{\theta}\right)+O\left(\frac{1}{n^2}\right), \tag{4.19}$$

which is equal to (4.18) up to the order $1/n$ as $n\to\infty$. It also follows from (4.15), (4.16) and (4.18) that the second-order asymptotic loss of $\hat{\gamma}_{ML^*}(=X_{(1)}^{**})$ relative to $\hat{\gamma}_{ML^*}^{\theta}(=X_{(1)}^{*})$ is given by

$$d_n\left(\hat{\gamma}_{ML^*},\ \hat{\gamma}_{ML^*}^{\theta}\right)=1+o(1) \tag{4.20}$$

as $n\to\infty$. Note that the loss is independent of θ up to the order $o(1)$. A similar consideration on the ratio $R_n(\hat{\gamma}_{ML^*},\hat{\gamma}_{ML^*}^{\theta})$ to Example 2.7.1 is done. From (4.17) and (4.19) it follows that the second-order asymptotic loss of $\hat{\gamma}_{UMVU}$ relative to $\hat{\gamma}_{UMVU}^{\theta}$ is

$$d_n\left(\hat{\gamma}_{UMVU}, \hat{\gamma}^{\theta}_{UMVU}\right) := n\left\{V_{\theta,\gamma}\left(\frac{n\theta}{\gamma}\hat{\gamma}_{UMVU}\right) - V_{\gamma}\left(\frac{n\theta}{\gamma}\hat{\gamma}^{\theta}_{UMVU}\right)\right\}$$
$$= \frac{n^2\theta}{(n-1)(n\theta-2)} = 1 + O\left(\frac{1}{n}\right),$$

which coincides with (4.20) up to the order $o(1)$ as $n \to \infty$.

Example 4.6.4 **(Lower-truncated beta distribution)** (Continued from Example 2.7.4). Let $c = 0$, $d = 1$, $a(x) = x^{-1}$, and $u(x) = \log x$ for $0 < \gamma \le x < 1$ in the density (1.7). Note that the density is uniform when $\theta = 1$. In Example 2.7.4, we have $b(\theta, \gamma) = \theta^{-1}(1 - \gamma^{\theta})$ for $\theta \in \Theta = (0, \infty)$ and the formulae of $\lambda_i(\theta, \gamma)$ $(i = 1, 2)$ and $\partial\lambda_1(\theta, \gamma)/\partial\gamma$. We also obtain

$$\lambda_3(\theta, \gamma) = -\frac{2}{\theta^3} - \frac{(\log\gamma)^3\gamma^{\theta}(1+\gamma^{\theta})}{(1-\gamma^{\theta})^3}, \quad k(\theta, \gamma) = \frac{\theta\gamma^{\theta-1}}{1-\gamma^{\theta}},$$
$$\frac{\partial k}{\partial\theta} = \frac{\gamma^{\theta-1}(1-\gamma^{\theta}+\theta\log\gamma)}{(1-\gamma^{\theta})^2},$$
$$\frac{\partial^2 k}{\partial\theta^2} = \frac{(\log\gamma)\gamma^{\theta-1}}{(1-\gamma^{\theta})^2}\{2(1-\gamma^{\theta}) + \theta\log\gamma\} + \frac{2(\log\gamma)\gamma^{2\theta-1}}{(1-\gamma^{\theta})^3}(1-\gamma^{\theta}+\theta\log\gamma),$$
$$\frac{\partial k}{\partial\gamma} = \frac{\theta\gamma^{\theta-1}(\theta-1+\gamma^{\theta-1})}{(1-\gamma^{\theta})^2},$$

hence, from (4.5), (4.6), and (4.8)

$$\hat{\gamma}^{\theta}_{ML*} = X^*_{(1)} = X_{(1)} - \frac{1 - X^{\theta}_{(1)}}{\theta X^{\theta-1}_{(1)} n},$$
$$T^*_{(1)} = T_{(1)} - \frac{1-\gamma^{\theta}}{\theta\gamma^{\theta-1}} + \frac{\theta - 1 + \gamma^{\theta-1}}{\theta\gamma^{\theta-1}n}T_{(1)} + O_p\left(\frac{1}{n^2}\right),$$
$$V_{\gamma}(kT^*_{(1)}) = 1 - \frac{2(\theta-1+\gamma^{\theta-1})}{\theta\gamma^{\theta-1}n} + O\left(\frac{1}{n^2}\right),$$

when θ is known. If θ is unknown, in a similar way to the above, from (4.11), (4.12) and (4.14) we can obtain the formulae of $\hat{\gamma}_{ML*}$ $(= X^{**}_{(1)})$, $T^{**}_{(1)}$ and $V_{\theta,\gamma}(kT^{**}_{(1)})$. Further, it follows from (4.15) that the second-order asymptotic loss of $\hat{\gamma}_{ML*}$ relative to $\hat{\gamma}^{\theta}_{ML*}$ is given by

$$d_n(\hat{\gamma}_{ML*}, \hat{\gamma}^{\theta}_{ML*}) = \frac{(1-\gamma^{\theta}+\theta\log\gamma)^2}{(1-\gamma^{\theta})^2 - \gamma^{\theta}(\theta\log\gamma)^2} + o(1)$$

as $n \to \infty$.

Example 4.6.5 **(Lower-truncated Erlang distribution)** (Continued from Example 2.7.5). Let $c = 0$, $d = \infty$, $a(x) = x^{j-1}$, and $u(x) = -x$ for $0 < \gamma \le x < \infty$ in the density (1.7), where $j = 1, 2, \ldots$. In Example 2.7.5, we have for each $j = 1, 2, \ldots, b_j(\theta, \gamma) = \int_\gamma^\infty x^{j-1} e^{-\theta x} dx$ for $\theta \in \Theta = (0, \infty)$ and the formulae of $\lambda_{ji} = (\partial^i/\partial\theta^i) \log b_j(\theta, \gamma) (i = 1, 2, 3)$, $\partial\lambda_{j1}(\theta, \gamma)/\partial\gamma$ and $k_j(\theta, \gamma)$. Let j be arbitrarily fixed in $\{1, 2, \ldots\}$. Since $\partial b_j/\partial\theta = -b_{j+1}$, it follows that

$$\frac{\partial k_j}{\partial \theta} = \frac{\gamma^{j-1} e^{-\theta\gamma}}{b_j^2}(b_{j+1} - \gamma b_j),$$

$$\frac{\partial^2 k_j}{\partial \theta^2} = \frac{\gamma^{j-1} e^{-\theta\gamma}}{b_j^3}\left\{b_j(\gamma^2 b_j - b_{j+2}) + 2b_{j+1}(b_{j+1} - \gamma b_j)\right\},$$

$$\frac{\partial k_j}{\partial \gamma} = \frac{\gamma^{j-2} e^{-\theta\gamma}}{b_j^2}\left\{(j - 1 - \theta\gamma)b_j + \gamma^j e^{-\theta\gamma}\right\},$$

where $k_j = k_j(\theta, \gamma)$ and $b_{j+i} = b_{j+i}(\theta, \gamma)$ $(i = 0, 1, 2)$, hence, from (4.5), (4.6) and (4.8)

$$\hat{\gamma}_{ML^*}^{\theta} = X_{(1)}^* = X_{(1)} - \frac{e^{\theta X_{(1)}} b_j(\theta, X_{(1)})}{X_{(1)}^{j-1} n},$$

$$T_{(1)}^* = T_{(1)} - \frac{e^{\theta\gamma} b_j}{\gamma^{j-1}} + \frac{e^{\theta\gamma} b_j}{\gamma^{j-1} n}\left\{\frac{e^{\theta\gamma} b_j(j - 1 - \theta\gamma)}{\gamma^j} - 1\right\} T_{(1)} + O_p\left(\frac{1}{n^2}\right),$$

$$V_\gamma(k_j T_{(1)}^*) = 1 - \frac{2e^{\theta\gamma} b_j}{\gamma^{j-1} n}\left\{\frac{e^{\theta\gamma} b_j(j - 1 - \theta\gamma)}{\gamma^j} - 1\right\} + O\left(\frac{1}{n^2}\right),$$

when θ is known. If θ is unknown, in a similar way to the above, from (4.11), (4.12) and (4.14) we can obtain the formulae of $\hat{\gamma}_{ML^*} (= X_{(1)}^{**})$, $T_{(1)}^{**}$ and $V_{\theta,\gamma}(kT_{(1)}^{**})$. Further, it follows from (4.15) that the asymptotic loss of $\hat{\gamma}_{ML^*}$ relative to $\hat{\gamma}_{ML^*}^{\theta}$ is given by

$$d_n(\hat{\gamma}_{ML^*}, \hat{\gamma}_{ML^*}^{\theta}) = \frac{1}{\lambda_{j2}}\left(\frac{b_{j+1}}{b_j} - \gamma\right)^2 + o(1)$$

as $n \to \infty$, where $\lambda_{j2} = (b_{j+2}/b_j) - (b_{j+1}/b_j)^2$.

A similar example to the above in the lower-truncated lognormal case in Example 2.7.6 is reduced to the lower-truncated normal case in Example 4.6.2.

4.7 Concluding Remarks

In a oTEF of distributions with a truncation parameter γ and a natural parameter θ, we considered the estimation problem of γ together with a bias-adjustment in the presence of a nuisance parameter θ. Using the stochastic expansions of the bias-adjusted MLEs $\hat{\gamma}_{ML^*}^{\theta}$ and $\hat{\gamma}_{ML^*}$ of γ when θ is known and when θ is unknown, respectively, we obtained their second-order asymptotic variances, from which the second-order asymptotic loss of $\hat{\gamma}_{ML^*}$ relative to $\hat{\gamma}_{ML^*}^{\theta}$ was derived. As is seen from Remark 4.5.1, the second-order asymptotic loss coincides with that of the bias-adjusted MLE $\hat{\theta}_{ML^*}$ of θ when γ is unknown relative to the MLE $\hat{\theta}_{ML}^{\gamma}$ of θ when γ is known, which means that the invariance on the second-order asymptotic loss holds even if the exchange of an interest parameter for a nuisance parameter is done. In the next chapter, the corresponding results to Theorems 4.3.1, 4.4.1 and 4.5.1 are obtained in the case of a two-sided truncated exponential family of distributions with two truncation parameters γ and ν and a natural parameter θ as a nuisance parameter, including an upper-truncated Pareto distribution which is important in applications.

4.8 Appendix C

Before proving Theorems 4.3.1 and 4.4.1, we prepare two lemmas.

Lemma 4.8.1 *It holds that*

$$E_{\theta,\gamma}(T_{(1)}^2) = \frac{2}{k^2(\theta,\gamma)} + \frac{6}{k(\theta,\gamma)n}A(\theta,\gamma) + O\left(\frac{1}{n^2}\right) \tag{4.21}$$

as $n \to \infty$, where $k(\theta,\gamma)$ and $A(\theta,\gamma)$ are given as (4.2) and (4.3), respectively.

The Eq. (4.21) can be obtained by straightforward calculation from Lemma 2.9.1.

Lemma 4.8.2 *Let $\hat{U} := \sqrt{\lambda_2 n}(\hat{\theta}_{ML} - \theta)$. Then, the asymptotic expectation of $\hat{U}$, $\hat{U}^2$, and $\hat{U}T_{(1)}$ are given by*

$$E_{\theta,\gamma}(\hat{U}) = -\frac{1}{\sqrt{\lambda_2 n}}\left\{\frac{1}{k}\left(\frac{\partial\lambda_1}{\partial\gamma}\right) + \frac{\lambda_3}{2\lambda_2}\right\} + O\left(\frac{1}{n\sqrt{n}}\right), \tag{4.22}$$

$$E_{\theta,\gamma}(\hat{U}^2) = 1 + O\left(\frac{1}{n}\right), \tag{4.23}$$

$$E_{\theta,\gamma}(\hat{U}T_{(1)}) = \frac{1}{k\sqrt{\lambda_2 n}}\left\{u(\gamma) - \lambda_1 - \frac{\lambda_3}{2\lambda_2}\right\} + O\left(\frac{1}{n\sqrt{n}}\right), \tag{4.24}$$

where $\lambda_j = \lambda_j(\theta,\gamma)$ $(j = 1, 2, 3)$ and $k = k(\theta,\gamma)$.

Proof The Eqs. (4.22) and (4.23) are given as (2.22) and (2.24), respectively, in the proof of Theorem 2.4.1. Since, by Theorem 2.4.1,

$$\hat{U} = Z_1 - \frac{\lambda_3}{2\lambda_2^{3/2}\sqrt{n}}Z_1^2 - \frac{1}{\sqrt{\lambda_2 n}}\left(\frac{\partial\lambda_1}{\partial\gamma}\right)T_{(1)} + O_p\left(\frac{1}{n}\right),$$

and, by Lemmas 2.9.2 and 2.9.3,

$$E_{\theta,\gamma}(Z_1 T_{(1)}) = \frac{1}{k\sqrt{\lambda_2 n}}\left\{u(\gamma) - \lambda_1 + \frac{2}{k}\left(\frac{\partial \lambda_1}{\partial \gamma}\right)\right\} + O\left(\frac{1}{n\sqrt{n}}\right),$$

$$E_{\theta,\gamma}(Z_1^2 T_{(1)}) = \frac{1}{k} + O\left(\frac{1}{n}\right), \tag{4.25}$$

it follows from (4.21) that

$$E_{\theta,\gamma}(\hat{U} T_{(1)}) = \frac{1}{k\sqrt{\lambda_2 n}}\left\{u(\gamma) - \lambda_1 - \frac{\lambda_3}{2\lambda_2}\right\} + O\left(\frac{1}{n\sqrt{n}}\right),$$

hence (4.24) is obtained. Thus, we complete the proof.

The proof of Theorem 4.3.1 By the Taylor expansion, we have from (4.2) and (4.3)

$$\hat{k}_\theta = k(\theta, X_{(1)}) = k(\theta, \gamma) + \frac{\partial k(\theta, \gamma)}{\partial \gamma} \cdot \frac{T_{(1)}}{n} + O_p\left(\frac{1}{n^2}\right), \tag{4.26}$$

$$\frac{\partial \hat{k}_\theta}{\partial \gamma} = \frac{\partial k}{\partial \gamma}(\theta, X_{(1)}) = \frac{\partial k(\theta, \gamma)}{\partial \gamma} + O_p\left(\frac{1}{n}\right), \tag{4.27}$$

$$\hat{A}_\theta = A(\theta, X_{(1)}) = A(\theta, \gamma) + O_p\left(\frac{1}{n}\right). \tag{4.28}$$

Since by (4.26)

$$\frac{1}{\hat{k}_\theta} = \frac{1}{k}\left\{1 - \frac{1}{k}\left(\frac{\partial k}{\partial \gamma}\right)\frac{T_{(1)}}{n} + O_p\left(\frac{1}{n^2}\right)\right\},$$

substituting (4.26)–(4.28) into (4.5), we obtain from (4.3)

$$\begin{aligned} T_{(1)}^* = n(X_{(1)}^* - \gamma) &= n(X_{(1)} - \gamma) - \frac{1}{\hat{k}_\theta} - \frac{1}{\hat{k}_\theta^3 n}\left(\frac{\partial \hat{k}_\theta}{\partial \gamma}\right) - \frac{1}{n}\hat{A}_\theta \\ &= T_{(1)} - \frac{1}{k} - \frac{1}{n}A + \frac{1}{kn}\left(\frac{\partial}{\partial \gamma}\log k\right)\left(T_{(1)} - \frac{1}{k}\right) + O_p\left(\frac{1}{n^2}\right) \\ &= T_{(1)} - \frac{1}{k} + \frac{1}{kn}\left(\frac{\partial}{\partial \gamma}\log k\right)T_{(1)} + O_p\left(\frac{1}{n^2}\right), \end{aligned}$$

where $k = k(\theta, \gamma)$ and $A = A(\theta, \gamma)$. Hence, we get (4.6). From (2.16), (4.3) and (4.6), it is easily seen that (4.7) holds, i.e., $E_\gamma(T_{(1)}^*) = O(1/n^2)$. From (2.16), (4.6) and (4.21), we have

$$E_\gamma\left[{T_{(1)}^*}^2\right] = \frac{1}{k^2} + \frac{4A}{kn} + \frac{2}{k^3 n}\left(\frac{\partial}{\partial \gamma}\log k\right) + O\left(\frac{1}{n^2}\right), \tag{4.29}$$

where $k = k(\theta, \gamma)$ and $A = A(\theta, \gamma)$. Hence, by (4.3) and (4.29)

$$E_\gamma\left[T_{(1)}^{*\,2}\right] = \frac{1}{k^2} - \frac{2}{k^3 n}\left(\frac{\partial}{\partial\gamma}\log k\right) + O\left(\frac{1}{n^2}\right). \tag{4.30}$$

From (4.7) and (4.30), we get (4.8). Thus, we complete the proof.

The proof of Theorem 4.4.1 By the Taylor expansion, we have

$$\begin{aligned}\hat{k} &= k\left(\hat{\theta}_{ML}, X_{(1)}\right) = k\left(\hat{\theta}_{ML}, \gamma + \frac{T_{(1)}}{n}\right)\\ &= k + \frac{1}{\sqrt{\lambda_2 n}}\left(\frac{\partial k}{\partial\theta}\right)\hat{U} + \frac{1}{n}\left(\frac{\partial k}{\partial\gamma}\right)T_{(1)} + \frac{1}{2\lambda_2 n}\left(\frac{\partial^2 k}{\partial\theta^2}\right)\hat{U}^2 + O_p\left(\frac{1}{n\sqrt{n}}\right).\end{aligned}$$

Since

$$\begin{aligned}\frac{1}{\hat{k}} &= \frac{1}{k} - \frac{1}{k^2\sqrt{\lambda_2 n}}\left(\frac{\partial k}{\partial\theta}\right)\hat{U} - \frac{1}{k^2 n}\left(\frac{\partial k}{\partial\gamma}\right)T_{(1)} - \frac{1}{2k^2\lambda_2 n}\left(\frac{\partial^2 k}{\partial\theta^2}\right)\hat{U}^2\\ &\qquad + \frac{1}{k^3\lambda_2 n}\left(\frac{\partial k}{\partial\theta}\right)^2\hat{U}^2 + O_p\left(\frac{1}{n\sqrt{n}}\right),\\ \hat{\lambda}_j &= \lambda_j\left(\hat{\theta}_{ML}, X_{(1)}\right) = \lambda_j(\theta, \gamma) + O_p\left(\frac{1}{\sqrt{n}}\right) \quad (j = 2, 3),\\ \frac{\partial^j\hat{k}}{\partial\theta^j} &= \frac{\partial^j k}{\partial\theta^j}\left(\hat{\theta}_{ML}, X_{(1)}\right) = \frac{\partial^j k}{\partial\theta^j}(\theta, \gamma) + O_p\left(\frac{1}{\sqrt{n}}\right) \quad (j = 1, 2),\\ \frac{\partial\hat{k}}{\partial\gamma} &= \frac{\partial k}{\partial\gamma}\left(\hat{\theta}_{ML}, X_{(1)}\right) = \frac{\partial k}{\partial\gamma}(\theta, \gamma) + O_p\left(\frac{1}{\sqrt{n}}\right),\\ \hat{A} &= A\left(\hat{\theta}_{ML}, X_{(1)}\right) = A(\theta, \gamma) + O_p\left(\frac{1}{\sqrt{n}}\right),\end{aligned}$$

it follows from (4.11) that

$$\begin{aligned}T_{(1)}^{**} &= n(X_{(1)}^{**} - \gamma)\\ &= T_{(1)} - \frac{1}{k} - \frac{1}{n}A + \frac{1}{k^2\sqrt{\lambda_2 n}}\left(\frac{\partial k}{\partial\theta}\right)\left\{\hat{U} + \frac{1}{\sqrt{\lambda_2 n}}\left(\frac{1}{k}\left(\frac{\partial\lambda_1}{\partial\gamma}\right) + \frac{\lambda_3}{2\lambda_2}\right)\right\}\\ &\quad + \frac{1}{k^2 n}\left(\frac{\partial k}{\partial\gamma}\right)\left(T_{(1)} - \frac{1}{k}\right) + \frac{1}{2k^2\lambda_2 n}\left[\frac{\partial^2 k}{\partial\theta^2} - \frac{2}{k}\left(\frac{\partial k}{\partial\theta}\right)^2\right]\left(\hat{U}^2 - 1\right)\\ &\qquad + O_p\left(\frac{1}{n\sqrt{n}}\right),\end{aligned} \tag{4.31}$$

where $k = k(\theta, \gamma)$, $A = A(\theta, \gamma)$, and $\lambda_j = \lambda_j(\theta, \gamma)$ $(j = 1, 2, 3)$, which derives (4.12) from (4.3). From (4.25), (4.31) and Lemmas 2.9.1, 4.8.1 and 4.8.2, we obtain (4.13) and

$$V_{\theta,\gamma}\left(T_{(1)}^{**}\right) = \frac{1}{k^2} + \frac{4A}{kn} + \frac{1}{k^4\lambda_2 n}\left(\frac{\partial k}{\partial \theta}\right)^2 + \frac{2}{k^4 n}\left(\frac{\partial k}{\partial \gamma}\right) + O\left(\frac{1}{n\sqrt{n}}\right). \quad (4.32)$$

Since

$$\frac{\partial}{\partial \theta}\log k(\theta,\gamma) = u(\gamma) - \lambda_1(\theta,\gamma),$$

it follows from (4.3) and (4.32) that

$$V_{\theta,\gamma}\left(T_{(1)}^{**}\right) = \frac{1}{k^2} - \frac{2}{k^3 n}\left(\frac{\partial}{\partial \gamma}\log k\right) + \frac{1}{k^2\lambda_2 n}(u(\gamma)-\lambda_1)^2 + O\left(\frac{1}{n\sqrt{n}}\right),$$

which shows that (4.14) holds. Thus, we complete the proof.

References

Akahira, M., & Ohyauchi, N. (2017). Second-order asymptotic loss of the MLE of a truncation parameter for a truncated exponential family of distributions. *Communication in Statistics: Theory and Methods*, *46*, 6085–6097.

Voinov, V. G. and Nikulin, M. S. (1993). *Unbiased estimators and their applications* (Vol. 1) Univariate Case. Dordrecht: Kluwer Academic Publishers.

Chapter 5
Estimation of a Truncation Parameter for a Two-Sided TEF

The corresponding results on maximum likelihood estimation of a truncation parameter together with a bias-adjustment to the case of oTEF in the previous chapter are obtained in the case of a two-sided truncated exponential family (tTEF) of distributions with two truncation parameters γ and ν and a natural parameter θ as a nuisance parameter.

5.1 Introduction

In this chapter, following mostly the paper by Akahira and Ohyauchi (2016), we obtain the corresponding results to the case of oTEF in the case of a tTEF of distributions with lower and upper truncation parameters γ and ν and a natural parameter θ as a nuisance parameter. In Sect. 5.3, we obtain a bias-adjusted MLE $\hat{\nu}_{ML^*}^{\theta,\gamma}$ of ν and derive its stochastic expansion and second-order asymptotic variance when θ and γ are known. In Sect. 5.4, we get a bias-adjusted MLE $\hat{\nu}_{ML^*}^{\gamma}$ of ν and derive its stochastic expansion and second-order asymptotic variance when θ is unknown and γ is known. In Sect. 5.5, we obtain a bias-adjusted MLE $\hat{\nu}_{ML^*}$ of ν and derive its stochastic expansion and second-order asymptotic variance when θ and γ are unknown. In Sect. 5.6, we get the second-order asymptotic losses of $\hat{\nu}_{ML^*}^{\gamma}$ and $\hat{\nu}_{ML^*}$ relative to $\hat{\nu}_{ML^*}^{\theta,\gamma}$ and in Sect. 5.7 give examples on a two-sided truncated exponential, a two-sided truncated normal, an upper-truncated Pareto, a two-sided truncated beta and a two-sided Erlang distributions. In particular, the results of Monte Carlo simulation discussed by Zhang (2013) from the viewpoint of minimum variance unbiased estimation in the upper-truncated Pareto case are theoretically clarified. Further, in Appendix D, the proofs of theorems are given.

M. Akahira, *Statistical Estimation for Truncated Exponential Families*,
JSS Research Series in Statistics, DOI 10.1007/978-981-10-5296-5_5

5.2 Preliminaries

According to Chap. 3, we have the formulation as follows. Suppose that $X_1, X_2, \dots,$ $X_n, \dots$ is a sequence of independent and identically distributed (i.i.d.) random variables according to $P_{\theta,\gamma,\nu}$ having the density (1.9), which belongs to a tTEF $\mathscr{P}_t$. Then, we consider the estimation problem on γ or ν in the presence of nuisance parameters θ and ν or γ, respectively. For any $\gamma, \nu \in (c, d)$, $\log b(\theta, \gamma, \nu)$ is strictly convex and infinitely differentiable in $\theta \in \Theta$ and

$$\lambda_j(\theta, \gamma, \nu) := \frac{\partial^j}{\partial \theta^j} \log b(\theta, \gamma, \nu) \tag{5.1}$$

is the j-th cumulant corresponding to (1.9) for $j = 1, 2, \cdots$. Let

$$\tilde{k}(\theta, \gamma, \nu) = a(\nu) e^{\theta u(\nu)} / b(\theta, \gamma, \nu), \tag{5.2}$$

$$\tilde{A}(\theta, \gamma, \nu) = \frac{1}{\tilde{k}^2(\theta, \gamma, \nu)} \left\{ \frac{\partial}{\partial \nu} \log \tilde{k}(\theta, \gamma, \nu) \right\}, \tag{5.3}$$

which are also defined in Theorem 3.4.1 and Lemma 3.9.1, respectively.

In the subsequent sections, we obtain the bias-adjusted MLE $\hat{\nu}_{ML^*}^{\theta,\gamma}$ of ν when θ and γ are known, the bias-adjusted MLE $\hat{\nu}_{ML^*}^{\gamma}$ of ν when θ is unknown and γ is known, and the bias-adjusted MLE $\hat{\nu}_{ML^*}$ of ν when θ and γ are unknown. Deriving their stochastic expansions and calculating their second-order asymptotic variances based on them, we get the second-order asymptotic losses of $\hat{\nu}_{ML^*}$ and $\hat{\nu}_{ML^*}^{\gamma}$ relative to $\hat{\nu}_{ML^*}^{\theta,\gamma}$.

5.3 Bias-Adjusted MLE $\hat{\nu}_{ML^*}^{\theta,\gamma}$ of ν When θ and γ are Known

For given $\boldsymbol{x} = (x_1, \dots, x_n)$ satisfying $c < \gamma \le x_{(1)} := \min_{1 \le i \le n} x_i$ and $x_{(n)} := \max_{1 \le i \le n} x_i \le \nu < d$, the likelihood function is given by

$$L^{\theta,\gamma}(\nu; \boldsymbol{x}) := \frac{1}{b^n(\theta, \gamma, \nu)} \left\{ \prod_{i=1}^n a(x_i) \right\} \exp\left\{ \theta \sum_{i=1}^n u(x_i) \right\},$$

when θ and γ are known. Then, the MLE $\hat{\nu}_{ML}^{\theta,\gamma}$ of ν is given by $X_{(n)} := \max_{1 \le i \le n} X_i$. Letting $T_{(n)} := n(X_{(n)} - \nu)$, we have the following.

Theorem 5.3.1 *For a tTEF $\mathscr{P}_t$ of distributions with densities of the form (1.9) with two truncation parameters γ and ν and a natural parameter θ, let $\hat{\nu}_{ML^*}^{\theta,\gamma} = X_{(n)}^*$ be a bias-adjusted MLE of ν such that*

$$X_{(n)}^* := X_{(n)} + \frac{1}{\hat{\tilde{k}}_{\theta,\gamma} n}, \tag{5.4}$$

where $\hat{\tilde{k}}_{\theta,\gamma} = \tilde{k}(\theta, \gamma, X_{(n)})$. *Then, the stochastic expansion of* $T_{(n)}^* := n(X_{(n)}^* - \nu)$ *is given by*

$$T_{(n)}^* = T_{(n)} + \frac{1}{\tilde{k}} - \frac{1}{\tilde{k}n}\left(\frac{\partial}{\partial \nu}\log \tilde{k}\right) T_{(n)} + O_p\left(\frac{1}{n^2}\right), \tag{5.5}$$

where $\tilde{k} = \tilde{k}(\theta, \gamma, \nu)$, *and the second-order asymptotic mean and asymptotic variance are given by*

$$E_\nu[T_{(n)}^*] = O\left(\frac{1}{n^2}\right), \tag{5.6}$$

$$V_\nu(\tilde{k}T_{(n)}^*) = 1 + \frac{2}{\tilde{k}n}\left(\frac{\partial}{\partial \nu}\log \tilde{k}\right) + O\left(\frac{1}{n^2}\right), \tag{5.7}$$

respectively.

5.4 Bias-Adjusted MLE $\hat{\nu}_{ML*}^{\gamma}$ of ν When θ is Unknown and γ is Known

In a similar way to Remark 4.5.3 in Chap. 4, we consider the case when θ is unknown and γ is known. Suppose that a random variable X has the density (1.9). Letting $Y = -X$, we have

$$f_Y(y;\theta,\delta,\eta) = \begin{cases} \frac{a_0(y)e^{\theta u_0(y)}}{b_{0\eta}(\theta,\delta)} & \text{for } -d < \delta \le y \le \eta < -c, \\ 0 & \text{otherwise} \end{cases} \tag{5.8}$$

as a density of Y, where $a_0(y) = a(-y)$, $u_0(y) = u(-y)$, $b_{0\eta}(\theta, \delta) = b(\theta, -\eta, -\delta)$, $\delta = -\nu$, and $\eta = -\gamma$. We put $Y_i = -X_i$ $(i = 1, 2, \cdots)$, $Y_{(1)} = -X_{(n)}$, $Y_{(n)} = -X_{(1)}$ and

$$\lambda_{\eta j}^0(\theta,\delta) = \frac{\partial^j}{\partial \theta^j}\log b_{0\eta}(\theta,\delta) \tag{5.9}$$

for $j = 1, 2, \ldots$. Since η is known, it is seen from (5.8) that the estimation problem on ν turns to that on δ in the oTEF of distributions which is treated in Chap. 4. Let $\hat{\delta}_{ML}^\eta = \hat{\delta}_{ML}^\eta(\boldsymbol{Y})$ and $\hat{\theta}_{ML}^\eta = \hat{\theta}_{ML}^\eta(\boldsymbol{Y})$ be the MLEs of δ and θ based on $\boldsymbol{Y} := (Y_1, \ldots, Y_n)$, respectively. From (5.8), it is seen that $\hat{\delta}_{ML}^\eta = Y_{(1)}$ and $\hat{\theta}_{ML}^\eta$ satisfy the likelihood equation

$$\frac{1}{n}\sum_{i=1}^{n} u_0(Y_i) - \lambda_{\eta 1}^0(\hat{\theta}_{ML}^{\eta}, Y_{(1)}) = 0,$$

that is, the MLE $\hat{\theta}_{ML}^{\gamma} = \hat{\theta}_{ML}^{\gamma}(\boldsymbol{X})$ of θ based on $\boldsymbol{X} := (X_1, \cdots, X_n)$ satisfies

$$\frac{1}{n}\sum_{i=1}^{n} u(X_i) - \lambda_1(\hat{\theta}_{ML}^{\gamma}, \gamma, X_{(n)}) = 0. \tag{5.10}$$

Let

$$k_{0\eta}(\theta, \delta) = \frac{a_0(\delta)e^{\theta u_0(\delta)}}{b_{0\eta}(\theta, \delta)}, \tag{5.11}$$

$$A_{0\eta}(\theta, \delta) := -\frac{1}{k_{0\eta}^2(\theta, \delta)}\left\{\frac{\partial}{\partial \delta}\log k_{0\eta}(\theta, \delta)\right\}. \tag{5.12}$$

Note that

$$\frac{\partial}{\partial \delta}\log k_{0\eta}(\theta, \delta) = \frac{a_0{}'(\delta)}{a_0(\delta)} + \theta u_0'(\delta) + k_{0\eta}(\theta, \delta). \tag{5.13}$$

Here, we have from (5.9) and (5.11)

$$k_{0\eta}(\theta, \delta) = \frac{a_0(\delta)e^{\theta u_0(\delta)}}{b_{0\eta}(\theta, \delta)} = \frac{a(\nu)e^{\theta u(\nu)}}{b(\theta, \gamma, \nu)} = \tilde{k}(\theta, \gamma, \nu) =: \tilde{k}_{\gamma}(\theta, \nu), \tag{5.14}$$

$$\begin{aligned}\lambda_{\eta j}^0(\theta, \delta) &= (\partial^j/\partial\theta^j)\log b_{0\eta}(\theta, \delta) = (\partial^j/\partial\theta^j)\log b(\theta, \gamma, \nu)\\ &= \lambda_j(\theta, \gamma, \nu) =: \lambda_{\gamma j}(\theta, \nu) \quad (j = 1, 2, \ldots).\end{aligned} \tag{5.15}$$

Let $T_{(1)} := n(Y_{(1)} - \delta)$, $k_{0\eta} = k_{0\eta}(\theta, \delta)$, $\lambda_{\eta j}^0 = \lambda_{\eta j}^0(\theta, \delta)$ $(j = 1, 2, \cdots)$, and $\hat{U}_{0\eta} = \sqrt{\lambda_{\eta 2}^0 n}(\hat{\theta}_{ML} - \theta)$. Then, we have the following.

Lemma 5.4.1 *For a tTEF of distributions having densities of the form (5.8) with truncation parameters δ and η and a natural parameter θ, let $\hat{\delta}_{ML*}^{\eta} = Y_{(1)}'$ be a bias-adjusted MLE of δ such that*

$$\begin{aligned}Y_{(1)}' = Y_{(1)} &- \frac{1}{\hat{k}_{0\eta}n} + \frac{1}{\hat{k}_{0\eta}^2\hat{\lambda}_{\eta 2}^0 n^2}\left(\frac{\partial \hat{k}_{0\eta}}{\partial\theta}\right)\left\{\frac{1}{\hat{k}_{0\eta}}\left(\frac{\partial\hat{\lambda}_{\eta 1}^0}{\partial\delta}\right) + \frac{\hat{\lambda}_{\eta 3}^0}{2\hat{\lambda}_{\eta 2}^0}\right\}\\ &- \frac{1}{2\hat{k}_{0\eta}^2\hat{\lambda}_{\eta 2}^0 n^2}\left\{\frac{\partial^2\hat{k}_{0\eta}}{\partial\theta^2} - \frac{2}{\hat{k}_{0\eta}}\left(\frac{\partial\hat{k}_{0\eta}}{\partial\theta}\right)^2\right\},\end{aligned} \tag{5.16}$$

where $\hat{k}_{0\eta} = k_{0\eta}(\hat{\theta}_{ML}, Y_{(1)})$, $\partial^j \hat{k}_{0\eta}/\partial\theta^j = (\partial^j/\partial\theta^j)k_{0\eta}(\hat{\theta}_{ML}, Y_{(1)})$ $(j = 1, 2)$, $\hat{\lambda}_{\eta j}^0 = \lambda_{\eta j}^0(\hat{\theta}_{ML}, Y_{(1)})(j = 2, 3)$, $\partial\hat{\lambda}_{\eta 1}^0/\partial\delta = (\partial/\partial\delta)\lambda_{\eta 1}^0(\hat{\theta}_{ML}, Y_{(1)})$. *Then, the stochastic expansion of* $T'_{(1)} := n(Y'_{(1)} - \delta)$ *is given by*

$$\begin{aligned} T'_{(1)} = T_{(1)} &- \frac{1}{k_{0\eta}} \\ &+ \frac{1}{k_{0\eta}^2\sqrt{\lambda_{\eta 2}^0 n}}\left(\frac{\partial k_{0\eta}}{\partial\theta}\right)\left[\hat{U}_{0\eta} + \frac{1}{\sqrt{\lambda_{\eta 2}^0 n}}\left\{\frac{1}{k_{0\eta}}\left(\frac{\partial\lambda_{\eta 1}^0}{\partial\delta}\right) + \frac{\lambda_{\eta 3}^0}{2\lambda_{\eta 2}^0}\right\}\right] \\ &+ \frac{1}{k_{0\eta}n}\left(\frac{\partial}{\partial\delta}\log k_{0\eta}\right)T_{(1)} + \frac{1}{2k_{0\eta}^2\lambda_{\eta 2}^0 n}\left\{\frac{\partial^2 k_{0\eta}}{\partial\theta^2} - \frac{2}{k_{0\eta}}\left(\frac{\partial k_{0\eta}}{\partial\theta}\right)^2\right\}\left(\hat{U}_{0\eta}^2 - 1\right) \\ &+ O_p\left(\frac{1}{n\sqrt{n}}\right), \end{aligned} \tag{5.17}$$

and the second-order asymptotic mean and asymptotic variance are given by

$$E_{\theta,\delta}\left[T'_{(1)}\right] = O\left(\frac{1}{n\sqrt{n}}\right), \tag{5.18}$$

$$\begin{aligned} V_{\theta,\delta}\left(k_{0\eta}T'_{(1)}\right) = 1 &- \frac{2}{k_{0\eta}n}\left(\frac{\partial}{\partial\delta}\log k_{0\eta}\right) + \frac{1}{\lambda_{\eta 2}^0 n}\left\{u_0(\delta) - \lambda_{\eta 1}^0\right\}^2 \\ &+ O\left(\frac{1}{n\sqrt{n}}\right), \end{aligned} \tag{5.19}$$

respectively.

The proof is omitted, since Lemma 5.4.1 is essentially same as Theorem 4.4.1. Let $\lambda_{\gamma 2}(\theta, \nu) := \lambda_2(\theta, \gamma, \nu)$ and $\hat{U}_\gamma = \sqrt{\lambda_{\gamma 2}n}(\hat{\theta}_{ML}^{\gamma} - \theta)$, where $\lambda_{\gamma 2} = \lambda_{\gamma 2}(\theta, \nu)$. Since $Y_{(1)} = -X_{(n)}$, from (5.11)–(5.16), we have the following.

Theorem 5.4.1 *For tTEF* $\mathscr{P}_t$ *of distributions having densities of the form (1.9) with truncation parameters* γ *and* ν *and a natural parameter* θ, *let* $\hat{\nu}_{ML*}^{\gamma} = X_{(n)}^{\dagger}$ *be a bias-adjusted MLE of* ν *such that*

$$\begin{aligned} X_{(n)}^{\dagger} = X_{(n)} &+ \frac{1}{\hat{\tilde{k}}_{\gamma}n} + \frac{1}{\hat{\tilde{k}}_{\gamma}^2\hat{\lambda}_{\gamma 2}n^2}\left(\frac{\partial\hat{\tilde{k}}_{\gamma}}{\partial\theta}\right)\left\{\frac{1}{\hat{\tilde{k}}_{\gamma}}\left(\frac{\partial\hat{\lambda}_{\gamma 1}}{\partial\nu}\right) - \frac{\hat{\lambda}_{\gamma 3}}{2\hat{\lambda}_{\gamma 2}}\right\} \\ &+ \frac{1}{2\hat{\tilde{k}}_{\gamma}^2\hat{\lambda}_{\gamma 2}n^2}\left\{\frac{\partial^2\hat{\tilde{k}}_{\gamma}}{\partial\theta^2} - \frac{2}{\hat{\tilde{k}}_{\gamma}}\left(\frac{\partial\hat{\tilde{k}}_{\gamma}}{\partial\theta}\right)^2\right\}, \end{aligned} \tag{5.20}$$

where

$$\begin{aligned}\hat{\tilde{k}}_\gamma &= \tilde{k}_\gamma(\hat{\theta}_{ML}, X_{(n)}) = \tilde{k}(\hat{\theta}_{ML}, \gamma, X_{(n)}),\\ \partial^j\hat{\tilde{k}}_\gamma/\partial\theta^j &= (\partial^j/\partial\theta^j)\tilde{k}_\gamma(\hat{\theta}_{ML}, X_{(n)}) = (\partial^j/\partial\theta^j)\tilde{k}(\hat{\theta}_{ML}, \gamma, X_{(n)}) \ (j=1,2),\\ \hat{\lambda}_{\gamma j} &= \lambda_{\gamma j}(\hat{\theta}_{ML}, X_{(n)}) = \lambda_j(\hat{\theta}_{ML}, \gamma, X_{(n)}) \ (j=2,3),\\ \partial\hat{\lambda}_{\gamma 1}/\partial\nu &= (\partial/\partial\nu)\lambda_{\gamma 1}(\hat{\theta}_{ML}, X_{(n)}) = (\partial/\partial\nu)\lambda_1(\hat{\theta}_{ML}, \gamma, X_{(n)}).\end{aligned}$$

Then, the stochastic expansion of $T^\dagger_{(n)} := n(X^\dagger_{(n)} - \nu)$ *is given by*

$$\begin{aligned}T^\dagger_{(n)} = T_{(n)} &+ \frac{1}{\tilde{k}_\gamma}\\ &- \frac{1}{\tilde{k}_\gamma^2\sqrt{\lambda_{\gamma 2}n}}\left(\frac{\partial\tilde{k}_\gamma}{\partial\theta}\right)\left[\hat{U}_\gamma - \frac{1}{\sqrt{\lambda_{\gamma 2}n}}\left\{\frac{1}{\tilde{k}_\gamma}\left(\frac{\partial\lambda_{\gamma 1}}{\partial\nu}\right) - \frac{\lambda_{\gamma 3}}{2\lambda_{\gamma 2}}\right\}\right]\\ &- \frac{1}{\tilde{k}_\gamma n}\left(\frac{\partial}{\partial\nu}\log\tilde{k}_\gamma\right)T_{(n)} - \frac{1}{2\tilde{k}_\gamma^2\lambda_{\gamma 2}n}\left\{\frac{\partial^2\tilde{k}_\gamma}{\partial\theta^2} - \frac{2}{\tilde{k}_\gamma}\left(\frac{\partial\tilde{k}_\gamma}{\partial\theta}\right)^2\right\}(\hat{U}_\gamma^2 - 1)\\ &+ O_p\left(\frac{1}{n\sqrt{n}}\right)\end{aligned} \tag{5.21}$$

where $\tilde{k}_\gamma = \tilde{k}_\gamma(\theta, \nu) = k(\theta, \gamma, \nu)$, $\lambda_{\gamma j} = \lambda_{\gamma j}(\theta, \nu) = \lambda_j(\theta, \gamma, \nu)$ $(j = 1, 2, 3)$, *and the second-order asymptotic mean and asymptotic variance are given by*

$$E_{\theta,\nu}[T^\dagger_{(n)}] = O\left(\frac{1}{n\sqrt{n}}\right), \tag{5.22}$$

$$V_{\theta,\nu}(\tilde{k}_\gamma T^\dagger_{(n)}) = 1 + \frac{2}{\tilde{k}_\gamma n}\left(\frac{\partial}{\partial\nu}\log\tilde{k}_\gamma\right) + \frac{1}{\lambda_{\gamma 2}n}\{u(\nu) - \lambda_{\gamma 1}\}^2 + O\left(\frac{1}{n\sqrt{n}}\right), \tag{5.23}$$

respectively.

5.5 Bias-Adjusted MLE $\hat{\nu}_{ML^*}$ of ν When θ and γ are Unknown

For given $\boldsymbol{x} = (x_1, \ldots, x_n)$ satisfying $c < \gamma \le x_{(1)}$ and $x_{(n)} \le \nu < d$, the likelihood function of θ, γ and ν is given by

$$L(\theta, \gamma, \nu; \boldsymbol{x}) := \frac{1}{b^n(\theta, \gamma, \nu)}\left\{\prod_{i=1}^n a(x_i)\right\}\exp\left\{\theta\sum_{i=1}^n u(x_i)\right\}. \tag{5.24}$$

Let $\hat{\theta}_{ML}$, $\hat{\gamma}_{ML}$, and $\hat{\nu}_{ML}$ be the MLEs of θ, γ, and ν, respectively. Then, it follows from (5.26) that $\hat{\gamma}_{ML} = X_{(1)}$, $\hat{\nu}_{ML} = X_{(n)}$, and $L(\hat{\theta}_{ML}, X_{(1)}, X_{(n)}; \boldsymbol{X}) = \sup_{\theta \in \Theta} L(\theta, X_{(1)}, X_{(n)}; \boldsymbol{X})$; hence, $\hat{\theta}_{ML}$ satisfies the likelihood equation

$$\frac{1}{n}\sum_{i=1}^{n} u(X_i) - \lambda_1(\hat{\theta}_{ML}, X_{(1)}, X_{(n)}) = 0,$$

where $\boldsymbol{X} := (X_1, \ldots, X_n)$. Let $\lambda_2 = \lambda_2(\theta, \gamma, \nu)$, $T_{(1)} := n(X_{(1)} - \gamma)$ and $\hat{U} := \sqrt{\lambda_2 n}(\hat{\theta}_{ML} - \theta)$. Then, we have the following.

Theorem 5.5.1 *For a tTEF $\mathscr{P}_t$ of distributions having densities of the form (1.9) with truncation parameters γ and ν and a natural parameter θ, let $\hat{\nu}_{ML^*} = X_{(n)}^{**}$ be a bias-adjusted MLE of ν such that*

$$X_{(n)}^{**} := X_{(n)} + \frac{1}{\hat{k}n} - \frac{1}{\hat{\tilde{k}}^2\hat{\lambda}_2 n^2}\left(\frac{\partial \hat{\tilde{k}}}{\partial \theta}\right)\left[\frac{1}{\hat{k}}\left(\frac{\partial \hat{\lambda}_1}{\partial \gamma}\right) - \frac{1}{\hat{\tilde{k}}}\left(\frac{\partial \hat{\lambda}_1}{\partial \nu}\right) + \frac{\hat{\lambda}_3}{2\hat{\lambda}_2}\right]$$
$$+ \frac{1}{\hat{\tilde{k}}^2\hat{k}n^2}\left(\frac{\partial \hat{\tilde{k}}}{\partial \gamma}\right) + \frac{1}{2\hat{\tilde{k}}^2\hat{\lambda}_2 n^2}\left\{\frac{\partial^2 \hat{\tilde{k}}}{\partial \theta^2} - \frac{2}{\hat{\tilde{k}}}\left(\frac{\partial \hat{\tilde{k}}}{\partial \theta}\right)^2\right\}, \tag{5.25}$$

where

$$\hat{\tilde{k}} = \tilde{k}(\hat{\theta}_{ML}, X_{(1)}, X_{(n)}), \quad \hat{k} = k(\hat{\theta}_{ML}, X_{(1)}, X_{(n)}),$$
$$\partial^j \hat{\tilde{k}}/\partial \theta^j = (\partial^j/\partial \theta^j)\tilde{k}(\hat{\theta}_{ML}, X_{(1)}, X_{(n)}) \ (j = 1, 2),$$
$$\partial \hat{\tilde{k}}/\partial \gamma = (\partial/\partial \gamma)\tilde{k}(\hat{\theta}_{ML}, X_{(1)}, X_{(n)}),$$
$$\partial \hat{\lambda}_1/\partial \gamma = (\partial/\partial \gamma)\lambda_1(\hat{\theta}_{ML}, X_{(1)}, X_{(n)}), \quad \partial \hat{\lambda}_1/\partial \nu = (\partial/\partial \nu)\lambda_1(\hat{\theta}_{ML}, X_{(1)}, X_{(n)}),$$
$$\hat{\lambda}_j = \lambda_j(\hat{\theta}_{ML}, X_{(1)}, X_{(n)}) \ (j = 2, 3).$$

Then, the stochastic expansion of $T_{(n)}^{**} := n(X_{(n)}^{**} - \nu)$ is given by

$$T_{(n)}^{**} = T_{(n)} + \frac{1}{\tilde{k}} - \frac{1}{\tilde{k}^2\sqrt{\lambda_2 n}}\left(\frac{\partial \tilde{k}}{\partial \theta}\right)\left[\hat{U} + \frac{1}{\sqrt{\lambda_2 n}}\left\{\frac{1}{k}\left(\frac{\partial \lambda_1}{\partial \gamma}\right) - \frac{1}{\tilde{k}}\left(\frac{\partial \lambda_1}{\partial \nu}\right) + \frac{\lambda_3}{2\lambda_2}\right\}\right]$$
$$- \frac{1}{\tilde{k}n}\left(\frac{\partial}{\partial \gamma}\log \tilde{k}\right)\left(T_{(1)} - \frac{1}{k}\right) - \frac{1}{\tilde{k}n}\left(\frac{\partial}{\partial \nu}\log \tilde{k}\right)T_{(n)}$$
$$- \frac{1}{2\tilde{k}^2\lambda_2 n}\left\{\frac{\partial^2 \tilde{k}}{\partial \theta^2} - \frac{2}{\tilde{k}}\left(\frac{\partial \tilde{k}}{\partial \theta}\right)^2\right\}(\hat{U}^2 - 1) + O_p\left(\frac{1}{n\sqrt{n}}\right), \tag{5.26}$$

where $\tilde{k} = \tilde{k}(\theta, \gamma, \nu)$, $\lambda_j = \lambda_j(\theta, \gamma, \nu)$ $(j = 1, 2, 3)$, *and the second-order asymptotic mean and asymptotic variance are given by*

$$E_{\theta,\gamma,\nu}[T^{**}_{(n)}] = O\left(\frac{1}{n\sqrt{n}}\right), \tag{5.27}$$

$$V_{\theta,\gamma,\nu}(\tilde{k}T^{**}_{(n)}) = 1 + \frac{2}{\tilde{k}n}\left(\frac{\partial}{\partial\nu}\log\tilde{k}\right) + \frac{1}{\lambda_2 n}\{u(\nu) - \lambda_1\}^2 + O\left(\frac{1}{n\sqrt{n}}\right), \tag{5.28}$$

respectively.

5.6 Second-Order Asymptotic Losses of $\hat{\nu}_{ML^*}$ and $\hat{\nu}^{\gamma}_{ML^*}$ Relative to $\hat{\nu}^{\theta,\gamma}_{ML^*}$

From the results in previous sections, we can asymptotically compare the bias-adjusted MLEs $\hat{\nu}^{\theta,\gamma}_{ML^*}$, $\hat{\nu}^{\gamma}_{ML^*}$, and $\hat{\nu}_{ML^*}$ of ν using their second-order asymptotic variances as follows.

Theorem 5.6.1 *For a tTEF $\mathscr{P}_t$ of distributions having densities of the form (1.9) with truncation parameters γ and ν and a natural parameter θ, let $\hat{\nu}^{\theta,\gamma}_{ML^*}$, $\hat{\nu}^{\gamma}_{ML^*}$, and $\hat{\nu}_{ML^*}$ be the bias-adjusted MLEs of ν when θ and γ are known, when θ is unknown and γ is known, and when θ and γ are unknown, respectively. Then, $\hat{\nu}_{ML^*} = X^{**}_{(n)}$ and $\hat{\nu}^{\gamma}_{ML^*} = X^{\dagger}_{(n)}$ are second-order asymptotically equivalent in the sense that*

$$d_n(\hat{\nu}_{ML^*}, \hat{\nu}^{\gamma}_{ML^*}) := n\{V_{\theta,\gamma,\nu}(\tilde{k}T^{**}_{(n)}) - V_{\theta,\gamma}(\tilde{k}_{\gamma}T^{\dagger}_{(n)})\} = o(1) \tag{5.29}$$

as $n \to \infty$. The second-order asymptotic losses of $\hat{\nu}_{ML^*}$ and $\hat{\nu}^{\gamma}_{ML^*}$ relative to $\hat{\nu}^{\theta,\gamma}_{ML^*} = X^{*}_{(n)}$ are given by

$$d_n(\hat{\nu}_{ML^*}, \hat{\nu}^{\theta,\gamma}_{ML^*}) := n\{V_{\theta,\gamma,\nu}(\tilde{k}T^{**}_{(n)}) - V_{\nu}(\tilde{k}T^{*}_{(n)})\} = \frac{\{u(\nu) - \lambda_1\}^2}{\lambda_2} + o(1) \tag{5.30}$$

$$d_n(\hat{\nu}^{\gamma}_{ML^*}, \hat{\nu}^{\theta,\gamma}_{ML^*}) := n\{V_{\theta,\nu}(\tilde{k}_{\gamma}T^{\dagger}_{(n)}) - V_{\nu}(\tilde{k}T^{*}_{(n)})\} = \frac{\{u(\nu) - \lambda_1\}^2}{\lambda_2} + o(1) \tag{5.31}$$

as $n \to \infty$, respectively.

The proof is straightforward from Theorems 5.3.1, 5.4.1, and 5.5.1, since $\tilde{k}_{\gamma} = \tilde{k}_{\gamma}(\theta, \nu) = \tilde{k}(\theta, \gamma, \nu) = \tilde{k}$ and $\lambda_{\gamma j} = \lambda_{\gamma j}(\theta, \nu) = \lambda_j(\theta, \gamma, \nu) = \lambda_j$ $(j = 1, 2)$.

Remark 5.6.1 It is seen from (1.6) and (5.30) that the ratio of the second-order asymptotic variance of $\tilde{k}T^{**}_{(n)}$ to that of $\tilde{k}T^{*}_{(n)}$ is given by

$$R_n(\hat{\nu}_{ML^*}, \hat{\nu}^{\theta,\gamma}_{ML^*}) = 1 + \frac{1}{\lambda_2 n}\{u(\nu) - \lambda_1\}^2 + o\left(\frac{1}{n}\right),$$

and similarly from (1.6) and (5.31)

$$R_n(\hat{\upsilon}^{\gamma}_{ML^*}, \hat{\upsilon}^{\theta,\gamma}_{ML^*}) = 1 + \frac{1}{\lambda_2 n}\{u(\nu) - \lambda_1\}^2 + o\left(\frac{1}{n}\right).$$

From the consideration of model in Sect. 1.1, using (1.5), (1.6), (5.30), and (5.31), we see that the difference between the asymptotic models $M(\hat{\upsilon}_{ML^*}, \theta, \gamma)$ and $M(\hat{\upsilon}^{\gamma}_{ML^*}, \theta, \gamma)$ is given by $d_n(\hat{\upsilon}_{ML^*}, \hat{\upsilon}^{\theta,\gamma}_{ML^*})$ or $R_n(\hat{\upsilon}_{ML^*}, \hat{\upsilon}^{\theta,\gamma}_{ML^*})$ up to the second order, through the MLE of ν. In a similar way to the above, the difference between $M(\hat{\upsilon}^{\gamma}_{ML^*}, \theta, \gamma)$ and $M(\hat{\upsilon}^{\theta,\gamma}_{ML^*}, \theta, \gamma)$ is given by $d_n(\hat{\upsilon}^{\gamma}_{ML^*}, \hat{\upsilon}^{\theta,\gamma}_{ML^*})$ or $R_n(\hat{\upsilon}^{\gamma}_{ML^*}, \hat{\upsilon}^{\theta,\gamma}_{ML^*})$ up to the second order.

5.7 Examples

Examples on the second-order asymptotic loss of the estimators are given in a two-sided truncated exponential, a two-sided truncated normal, an upper-truncated Pareto, a two-sided truncated beta, and a two-sided truncated Erlang cases, which are treated in Chap. 3.

Example 5.7.1 **(Two-sided truncated exponential distribution)** (Continued from Example 3.7.1). Let $c = -\infty$, $d = \infty$, $a(x) \equiv 1$, and $u(x) = -x$ for $-\infty < \gamma \le x \le \nu < \infty$ in the density (1.9). Since $b(\theta, \gamma, \nu) = (e^{-\theta\gamma} - e^{-\theta\nu})/\theta$ for $\theta \in \Theta = (0, \infty)$, it follows from (5.1) that

$$\lambda_1 = \frac{\partial}{\partial\theta}\log b(\theta, \gamma, \nu) = \frac{-\gamma e^{-\theta\gamma} + \nu e^{-\theta\nu}}{e^{-\theta\gamma} - e^{-\theta\nu}} - \frac{1}{\theta} = \frac{-\gamma e^{-\xi} + \nu}{e^{-\xi} - 1} - \frac{1}{\theta}, \tag{5.32}$$

$$\begin{aligned}\lambda_2 = \frac{\partial^2}{\partial\theta^2}\log b(\theta, \gamma, \nu) &= \frac{\gamma^2 e^{-\theta\gamma} - \nu^2 e^{-\theta\nu}}{e^{-\theta\gamma} - e^{-\theta\nu}} - \frac{(\gamma e^{-\theta\gamma} - \nu e^{-\theta\nu})^2}{(e^{-\theta\gamma} - e^{-\theta\nu})^2} + \frac{1}{\theta^2} \\ &= \frac{(e^{-\xi} - 1)^2 - \xi^2 e^{-\xi}}{\theta^2(e^{-\xi} - 1)^2},\end{aligned} \tag{5.33}$$

$$\tilde{k}(\theta, \gamma, \nu) = \frac{\theta e^{-\theta\nu}}{e^{-\theta\gamma} - e^{-\theta\nu}} = \frac{\theta}{e^{-\xi} - 1}, \tag{5.34}$$

where $\xi = \theta(\gamma - \nu)$. Hence, by (5.4) and (5.34), the bias-adjusted MLE $\hat{\upsilon}^{\theta,\gamma}_{ML^*}$ of ν is

$$X^*_{(n)} = X_{(n)} + \frac{1}{\theta n}\left\{e^{\theta(X_{(n)} - \gamma)} - 1\right\},$$

when θ and γ are known. It also follows from (5.7) and (5.34) that

$$V_\nu(\tilde{k}T^*_{(n)}) = 1 - \frac{2}{n}e^{\theta(\nu - \gamma)} + O\left(\frac{1}{n^2}\right).$$

Next, we consider the case when θ is unknown and γ is known. Then, it follows from (5.10) that the MLE $\hat{\theta}^{\gamma}_{ML}$ satisfies the likelihood equation

$$\bar{X} = \frac{\gamma e^{\hat{\theta}^{\gamma}_{ML}(X_{(n)}-\gamma)} - X_{(n)}}{e^{\hat{\theta}^{\gamma}_{ML}(X_{(n)}-\gamma)} - 1} + \frac{1}{\hat{\theta}^{\gamma}_{ML}},$$

where $\bar{X} := (1/n)\sum_{i=1}^{n} X_i$. Since

$$\frac{\partial \tilde{k}}{\partial \theta}(\theta,\gamma,\nu) = \frac{(1+\xi)e^{-\xi}-1}{(e^{-\xi}-1)^2}, \quad \frac{\partial^2 \tilde{k}}{\partial \theta^2}(\theta,\gamma,\nu) = \frac{\xi e^{-\xi}}{\theta(e^{-\xi}-1)^3}\{(\xi+2)e^{-\xi}+\xi-2\}$$

$$\lambda_3(\theta,\gamma,\nu) = \frac{\partial \lambda_2}{\partial \theta}(\theta,\gamma,\nu) = -\frac{1}{\theta^3}\left\{2+\frac{\xi^3 e^{-\xi}(e^{-\xi}+1)}{(e^{-\xi}-1)^3}\right\},$$

$$\frac{\partial \lambda_1}{\partial \nu}(\theta,\gamma,\nu) = \frac{(1+\xi)e^{-\xi}-1}{(e^{-\xi}-1)^2},$$

it follows from (5.20) and (5.32)–(5.34) that the bias-adjusted MLE $\hat{\nu}^{\gamma}_{ML^*} = X^{\dagger}_{(n)}$ is obtained. When θ and γ are unknown, the MLE $\hat{\theta}_{ML}$ of θ satisfies the likelihood equation

$$\bar{X} = \frac{X_{(1)}e^{\hat{\theta}_{ML}(X_{(n)}-X_{(1)})} - X_{(n)}}{e^{\hat{\theta}_{ML}(X_{(n)}-X_{(1)})} - 1} + \frac{1}{\hat{\theta}_{ML}}.$$

In a similar way to the above, the bias-adjusted MLE $\hat{\nu}_{ML^*} = X^{**}_{(n)}$ is obtained from (5.25) and (5.32)–(5.34). Hence, it follows from (5.29)–(5.31) that $d_n(\hat{\nu}_{ML^*}, \hat{\nu}^{\gamma}_{ML^*}) = o(1)$ and

$$d_n(\hat{\nu}_{ML^*}, \hat{\nu}^{\theta,\gamma}_{ML^*}) = d_n(\hat{\nu}^{\gamma}_{ML^*}, \hat{\nu}^{\theta,\gamma}_{ML^*}) = \frac{\{(1+\xi)e^{-\xi}-1\}^2}{(e^{-\xi}-1)^2-\xi^2 e^{-\xi}} + o(1)$$

as $n \to \infty$. When $\theta = 1$ and $\gamma - \nu = -1, -2, -3$, the values of second-order asymptotic loss $d_n(\hat{\nu}_{ML^*}, \hat{\nu}^{\theta,\gamma}_{ML^*}) = d_n(\hat{\nu}^{\gamma}_{ML^*}, \hat{\nu}^{\theta,\gamma}_{ML^*})$ and the ratio $R_n(\hat{\nu}_{ML^*}, \hat{\nu}^{\theta,\gamma}_{ML^*}) = R_n(\hat{\nu}^{\gamma}_{ML^*}, \hat{\nu}^{\theta,\gamma}_{ML^*})$ up to the order $1/n$ are obtained from the above and Remark 5.6.1 (see Table 5.1 and Fig. 5.1).

It is noted from (5.3) and (3.17) in Appendix B1 that the second-order asymptotic mean of $T_{(n)} = n(X_{(n)} - \nu)$ is given by

$$E_{\theta,\gamma,\nu}(T_{(n)}) = -\frac{1}{\theta}\left\{e^{\theta(\nu-\gamma)}-1\right\} + \frac{1}{\theta n}e^{\theta(\nu-\gamma)}\left\{e^{\theta(\nu-\gamma)}-1\right\} + O\left(\frac{1}{n^2}\right).$$

Example 5.7.2 (**Two-sided truncated normal distribution**) (Continued from Example 3.7.2). Let $c = -\infty$, $d = \infty$, $a(x) = e^{-x^2/2}$, and $u(x) = x$ for $-\infty < \gamma \le x \le \nu < \infty$ in the density (1.9). Since

Table 5.1 Values of $d_n(\hat{\nu}_{ML^*}, \hat{\nu}_{ML^*}^{\theta,\gamma})$ and $R_n(\hat{\nu}_{ML^*}, \hat{\nu}_{ML^*}^{\theta,\gamma})$ for $\theta = 1$ and $\gamma - \nu = -1, \ -2, \ -3$

$\xi = \gamma - \nu$	$d_n(\hat{\nu}_{ML^*}, \hat{\nu}_{ML^*}^{\theta,\gamma})$	$R_n(\hat{\nu}_{ML^*}, \hat{\nu}_{ML^*}^{\theta,\gamma})$
-1	$4.2699 + o(1)$	$1 + \dfrac{4.2699}{n} + o\left(\dfrac{1}{n}\right)$
-2	$6.2480 + o(1)$	$1 + \dfrac{6.2480}{n} + o\left(\dfrac{1}{n}\right)$
-3	$9.2380 + o(1)$	$1 + \dfrac{9.2380}{n} + o\left(\dfrac{1}{n}\right)$

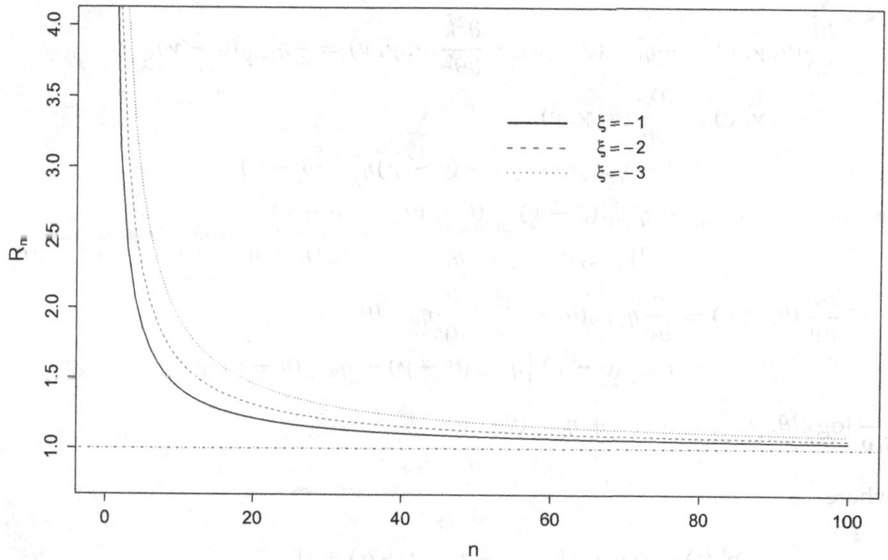

Fig. 5.1 Graph of the ratio $R_n(\hat{\nu}_{ML^*}, \hat{\nu}_{ML^*}^{\theta,\gamma})$ up to the order $1/n$ for $\theta = 1$ and $\xi = \gamma - \nu = -1, \ -2, \ -3$

$$b(\theta, \gamma, \nu) = \sqrt{2\pi} e^{\theta^2/2}\{\Phi(\theta - \gamma) - \Phi(\theta - \nu)\}$$

for $\theta \in \Theta = (-\infty, \infty)$, it follows that

$$\begin{aligned}
\lambda_1(\theta, \gamma, \nu) &= \theta + \eta_{\gamma-\nu}(\theta - \gamma) + \eta_{\nu-\gamma}(\theta - \nu), \\
\lambda_2(\theta, \gamma, \nu) &= 1 - (\theta - \gamma)\eta_{\gamma-\nu}(\theta - \gamma) - (\theta - \nu)\eta_{\nu-\gamma}(\theta - \nu) \\
&\quad - \{\eta_{\gamma-\nu}(\theta - \gamma) + \eta_{\nu-\gamma}(\theta - \nu)\}^2, \\
\tilde{k}(\theta, \gamma, \nu) &= -\eta_{\nu-\gamma}(\theta - \nu),
\end{aligned}$$

where $\eta_\alpha(t) := \phi(t)/\{\Phi(t) - \Phi(t + \alpha)\}$ with $\Phi(t) = \int_{-\infty}^{t} \phi(x)dx$ and $\phi(x) = (1/\sqrt{2\pi})e^{-x^2/2}$ for $-\infty < x < \infty$. When θ and γ are known, it follows from (5.4) that the bias-adjusted MLE $\hat{\nu}_{ML^*}^{\theta,\gamma}$ of ν is

$$X_{(n)}^* = X_{(n)} - \frac{1}{\eta_{X_{(n)}-\gamma}(\theta - X_{(n)})n}.$$

Next, we consider the case when θ is unknown and γ is known. Then, it follows from (5.10) that the MLE $\hat{\theta}_{ML}^{\gamma}$ of θ satisfies the likelihood equation

$$\bar{X} - \hat{\theta}_{ML}^{\gamma} - \eta_{\gamma-X_{(n)}}(\hat{\theta}_{ML}^{\gamma} - \gamma) - \eta_{X_{(n)}-\gamma}(\hat{\theta}_{ML}^{\gamma} - X_{(n)}) = 0.$$

Since

$$\begin{aligned}
\frac{\partial \tilde{k}}{\partial \theta}(\theta, \gamma, \nu) &= -\eta_{\nu-\gamma}'(\theta - \gamma), \quad \frac{\partial^2 \tilde{k}}{\partial \theta^2}(\theta, \gamma, \nu) = -\eta_{\nu-\gamma}''(\theta - \gamma), \\
\lambda_3(\theta, \gamma, \nu) &= \frac{\partial \lambda_2}{\partial \theta}(\theta, \gamma, \nu) \\
&= 1 - \eta_{\gamma-\nu}(\theta - \gamma) - (\theta - \gamma)\eta_{\gamma-\nu}'(\theta - \gamma) \\
&\quad - \eta_{\nu-\gamma}(\theta - \nu) - (\theta - \nu)\eta_{\nu-\gamma}'(\theta - \nu) \\
&\quad - 2\{\eta_{\gamma-\nu}(\theta - \gamma) + \eta_{\nu-\gamma}(\theta - \nu)\}\{\eta_{\gamma-\nu}'(\theta - \gamma) + \eta_{\nu-\gamma}'(\theta - \nu)\}, \\
\frac{\partial \lambda_1}{\partial \nu}(\theta, \gamma, \nu) &= \frac{\partial}{\partial \nu}\eta_{\gamma-\nu}(\theta - \gamma) + \frac{\partial}{\partial \nu}\eta_{\nu-\gamma}(\theta - \nu) \\
&= \eta_{\nu-\gamma}(\theta - \nu)\left\{\eta_{\gamma-\nu}(\theta - \gamma) + \eta_{\nu-\gamma}(\theta - \nu) + \theta - \nu\right\}, \\
\frac{\partial}{\partial \nu}\log \tilde{k}(\theta, \gamma, \nu) &= \theta - \nu + \eta_{\nu-\gamma}(\theta - \nu),
\end{aligned}$$

where

$$\begin{aligned}
\eta_{\alpha}'(t) &= -\eta_{\alpha}(t)\left\{\eta_{\alpha}(t) + \eta_{-\alpha}(t + \alpha) + t\right\}, \\
\eta_{\alpha}''(t) &= \eta_{\alpha}(t)\left\{t\eta_{\alpha}(t) + (t - \alpha)\eta_{-\alpha}(t + \alpha) + t^2 + 1\right\},
\end{aligned}$$

it follows from (5.20) that the bias-adjusted MLE $\hat{\nu}_{ML^*}^{\gamma} = X_{(n)}^{\dagger}$ is obtained. When θ and γ are unknown, the MLE $\hat{\theta}_{ML}$ of θ satisfies the likelihood equation

$$\bar{X} - \hat{\theta}_{ML} - \eta_{X_{(1)}-X_{(n)}}(\hat{\theta}_{ML} - X_{(1)}) - \eta_{X_{(n)}-X_{(1)}}(\hat{\theta}_{ML} - X_{(n)}) = 0.$$

In a similar way to the above, the bias-adjusted MLE $\hat{\nu}_{ML^*} = X_{(n)}^{**}$ is obtained from (5.25). Hence, it follows from (5.29)–(5.31) that $d_n(\hat{\nu}_{ML^*}, \hat{\nu}_{ML^*}^{\gamma}) = o(1)$ and

$$\begin{aligned}
&d_n(\hat{\nu}_{ML^*}, \hat{\nu}_{ML^*}^{\theta,\gamma}) = d_n(\hat{\nu}_{ML^*}^{\gamma}, \hat{\nu}_{ML^*}^{\theta,\gamma}) \\
&= \frac{\{\theta - \nu + \eta_{\gamma-\nu}(\theta - \gamma) + \eta_{\nu-\gamma}(\theta - \nu)\}^2}{1 - (\theta - \gamma)\eta_{\gamma-\nu}(\theta - \gamma) - (\theta - \nu)\eta_{\nu-\gamma}(\theta - \nu) - \{\eta_{\gamma-\nu}(\theta - \gamma) + \eta_{\nu-\gamma}(\theta - \nu)\}^2} \\
&\qquad + o(1)
\end{aligned}$$

as $n \to \infty$. It is also noted that the second-order asymptotic mean of $T_{(n)} = n(X_{(n)} - \nu)$ is given by

$$E_{\theta,\gamma,\nu}(T_{(n)}) = \frac{1}{\eta_{\nu-\gamma}(\theta - \nu)} - \frac{1}{\eta_{\nu-\gamma}^2(\theta - \nu)n}\{\theta - \nu + \eta_{\nu-\gamma}(\theta - \nu)\} + O\left(\frac{1}{n^2}\right).$$

Example 5.7.3 (**Upper-truncated Pareto distribution**) (Continued from Example 3.7.3). Let $c = 0$, $d = \infty$, $a(x) = 1/x$, and $u(x) = -\log x$ for $0 < \gamma \le x \le \nu < \infty$ in the density (1.9), which yields the upper-truncated Pareto distribution. Then, $b(\theta, \gamma, \nu) = \{1 - (\gamma/\nu)^\theta\}/(\theta\gamma^\theta)$ for $\theta \in \Theta = (0, \infty)$, and

$$\tilde{k}(\theta, \gamma, \nu) = \frac{\theta/\nu}{(\nu/\gamma)^\theta - 1}. \tag{5.35}$$

For the upper-truncated Pareto distribution with an index parameter θ and truncation parameters γ and ν, Zhang (2013) obtained the asymptotic biases of the MLEs $\hat{\gamma}_{ML} = X_{(1)}$ and $\hat{\nu}_{ML} = X_{(n)}$ of γ and ν and showed that the UMVU estimator of γ was

$$\hat{\gamma}_{UMVU}^{\theta,\nu} = X_{(1)}\left[1 + \frac{1}{\theta n}\left\{\left(\frac{X_{(1)}}{\nu}\right)^\theta - 1\right\}\right],$$

when θ and ν were known and the UMVU estimator of ν was

$$\hat{\nu}_{UMVU}^{\theta,\gamma} = X_{(n)}\left[1 + \frac{1}{\theta n}\left\{\left(\frac{X_{(n)}}{\gamma}\right)^\theta - 1\right\}\right], \tag{5.36}$$

when θ and γ were known. Note that (θ, γ, ν) is presented as (α, β, γ) in the paper by Zhang (2013). Put $\xi = (\gamma/\nu)^\theta$. Letting $t = \log x$, $\gamma_0 = \log \gamma$, and $\nu_0 = \log \nu$, we see that the density of upper-truncated Pareto distribution becomes

$$f(t; \theta, \gamma_0, \nu_0) = \begin{cases} \frac{\theta e^{\theta\gamma_0}}{1 - e^{-\theta(\nu_0 - \gamma_0)}} e^{-\theta t} & \text{for } \gamma_0 \le t \le \nu_0, \\ 0 & \text{otherwise.} \end{cases}$$

Here, note that $\xi_0 := \theta(\gamma_0 - \nu_0) = \log \xi$. Hence, the upper-truncated Pareto case is reduced to the two-sided truncated exponential one in Example 3.7.1. Replacing $\bar{X}$, $X_{(1)}$, and $X_{(n)}$ by $\overline{\log X} := (1/n)\sum_{i=1}^n \log X_i$, $\log X_{(1)}$, $\log X_{(n)}$, respectively, in Example 5.7.1, we have the second-order asymptotic losses

$$d_n(\hat{\nu}_{ML^*}, \hat{\nu}_{ML^*}^{\gamma}) = o(1), \quad d_n(\hat{\nu}_{ML^*}, \hat{\nu}_{ML^*}^{\theta,\gamma}) = \frac{(1 - \xi + \log \xi)^2}{(1 - \xi)^2 - \xi(\log \xi)^2} + o(1) \tag{5.37}$$

Table 5.2 Values of $d_n(\hat{\upsilon}_{ML^*}, \hat{\upsilon}_{ML^*}^{\theta,\gamma})$ and $R_n(\hat{\upsilon}_{ML^*}, \hat{\upsilon}_{ML^*}^{\theta,\gamma})$ for $\theta = \gamma = 1$ and $\nu = 2,\ 3,\ 5$

ξ	$d_n(\hat{\upsilon}_{ML^*}, \hat{\upsilon}_{ML^*}^{\theta,\gamma})$	$R_n(\hat{\upsilon}_{ML^*}, \hat{\upsilon}_{ML^*}^{\theta,\gamma})$
1/2	$3.8170 + o(1)$	$1 + \frac{3.8170}{n} + o\left(\frac{1}{n}\right)$
1/3	$4.4288 + o(1)$	$1 + \frac{4.4288}{n} + o\left(\frac{1}{n}\right)$
1/5	$5.3730 + o(1)$	$1 + \frac{5.3730}{n} + o\left(\frac{1}{n}\right)$

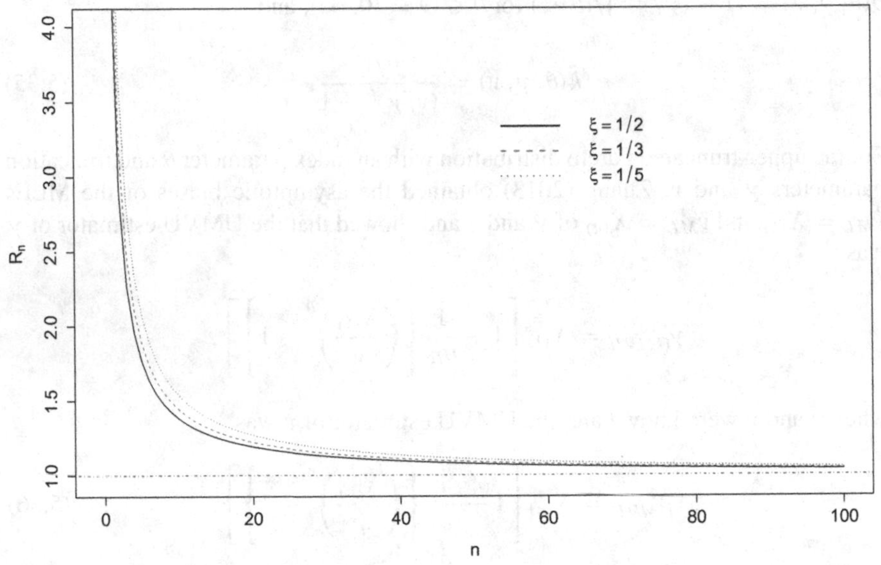

Fig. 5.2 Graph of the ratio $R_n(\hat{\upsilon}_{ML^*}, \hat{\upsilon}_{ML^*}^{\theta,\gamma})$ up to the order $1/n$ for $\theta = \gamma = 1$ and $\nu = 2,\ 3,\ 5$

as $n \to \infty$. When $\theta = \gamma = 1$ and $\nu = 2,\ 3,\ 5$, the values of second-order asymptotic loss $d_n(\hat{\upsilon}_{ML^*}, \hat{\upsilon}_{ML^*}^{\theta,\gamma})$ and ratio $R_n(\hat{\upsilon}_{ML^*}, \hat{\upsilon}_{ML^*}^{\theta,\gamma})$ up to the order $1/n$ are obtained from the above and Remark 5.6.1 (see Table 5.2 and Fig. 5.2).

It follows from (5.4) and (5.36) that

$$X_{(n)}^* = X_{(n)}\left[1 + \frac{1}{\theta n}\left\{\left(\frac{X_{(n)}}{\gamma}\right)^{\theta} - 1\right\}\right] = \hat{\upsilon}_{UMVU}^{\theta,\gamma} \tag{5.38}$$

and for the bias-adjusted MLE $\hat{\upsilon}_{ML^*}^{\theta,\gamma} = X_{(n)}^*$

$$\begin{aligned} T_{(n)}^* &= n(X_{(n)}^* - \nu) \\ &= T_{(n)} + \frac{\nu}{\theta}\left\{\left(\frac{\nu}{\gamma}\right)^{\theta} - 1\right\} + \frac{1}{n}\left\{\left(1 + \frac{1}{\theta}\right)\left(\frac{\nu}{\gamma}\right)^{\theta} - \frac{1}{\theta}\right\} T_{(n)} + O_p\left(\frac{1}{n^2}\right). \end{aligned}$$

Since, by (5.38)

$$T^*_{(n)} = n(\hat{\nu}^{\theta,\gamma}_{UMVU} - \nu),$$

when θ and γ are known, it follows from (5.7) that

$$V_\nu(\tilde{k}T^*_{(n)}) = V_\nu(\tilde{k}n(\hat{\nu}^{\theta,\gamma}_{UMVU} - \nu)) = 1 - \frac{2}{\theta n}\left\{(\theta+1)\left(\frac{\nu}{\gamma}\right)^\theta - 1\right\} + O\left(\frac{1}{n^2}\right). \tag{5.39}$$

From the result of Monte Carlo simulation, Zhang (2013) mentioned that, for example, when both θ and ν/γ were large, say $(\theta, \gamma, \nu) = (5, 1, 7)$, the variance of $\hat{\nu}^{\theta,\gamma}_{UMVU}$ could be very large for finite sample sizes. Indeed, from (5.39), we have

$$V_\nu(\hat{\nu}^{\theta,\gamma}_{UMVU}) = \frac{1}{\tilde{k}^2 n^2} + O\left(\frac{1}{n^3}\right). \tag{5.40}$$

For $(\theta, \gamma, \nu) = (5, 1, 7)$, we obtain from (5.35)

$$\frac{1}{\tilde{k}} = \frac{7^5 - 1}{5/7} \fallingdotseq 23528.4,$$

hence, for fixed n, the first term is seen to be very large in the order of $1/n^2$ in the right-hand side of (5.40). But, in order to compare estimators in terms of variance, we need a standardization such a form as (5.39). From the result of Monte Carlo simulation, Zhang (2013) also stated that for fixed γ, $\hat{\nu}^{\theta,\gamma}_{UMVU}$ behaved better and better when θ went to 0. Indeed, in this case, it is easily seen from (5.35) that $\tilde{k}(\theta, \gamma, \nu) \to \infty$ as $\theta \to 0$; hence from (5.40), the variance of $\hat{\nu}^{\theta,\gamma}_{UMVU}$ becomes very small. Further, Zhang (2013) considered plug-in estimators $\hat{\gamma}^{\hat{\theta}_{ML}, X_{(n)}}_{UMVU}$ and $\hat{\nu}^{0,X_{(1)}}_{UMVU}$ using the MLE $\hat{\theta}_{ML}$ of θ and the MLEs $X_{(1)}$ and $X_{(n)}$ of γ and ν, respectively, when θ, γ, and ν are unknown, and from the result of Monte Carlo simulation concluded that the improvement of $\hat{\nu}^{0,X_{(1)}}_{UMVU}$ was significant only if θ was small, but its poor behavior for large θ was to be expected. First, we consider a plug-in estimator $\nu^{\hat{\theta}_{ML}, X_{(1)}}_{UMVU}$ using the MLEs $\hat{\theta}_{ML}$ and $X_{(1)}$ of θ and γ, respectively, when θ, γ, and ν are unknown. It is noted that $\hat{\theta}_{ML}$ satisfies the likelihood equation

$$\frac{1}{\hat{\theta}_{ML}} + \frac{(X_{(1)}/X_{(n)})^{\hat{\theta}_{ML}} \log(X_{(1)}/X_{(n)})}{1 - (X_{(1)}/X_{(n)})^{\hat{\theta}_{ML}}} = \frac{1}{n}\sum_{i=1}^n \log\frac{X_{(i)}}{X_{(1)}}.$$

Then, we have

$$\begin{aligned}
&E_{\theta,\gamma,\nu}[n(\hat{\nu}_{UMVU}^{\hat{\theta}_{ML},X_{(1)}} - \nu)] \\
&= \frac{\nu}{\theta n} - \frac{\nu}{\theta^3 \xi \lambda_2 n}\left\{(1-\xi+\log\xi)\left(1+\frac{1+\xi}{1-\xi}\log\xi - \frac{\theta\lambda_3}{2\lambda_2}\right) - \frac{1}{2}(\log\xi)^2\right\} \\
&\quad + O\left(\frac{1}{n\sqrt{n}}\right),
\end{aligned} \tag{5.41}$$

which implies that it is asymptotically unbiased up to the order $o(1)$, but not up to the order $o(1/n)$, where

$$\lambda_2 = \frac{1}{\theta^2}\left\{1 - \frac{\xi(\log\xi)^2}{(1-\xi)^2}\right\}, \quad \lambda_3 = -\frac{2}{\theta^3}\left\{1 + \frac{\xi(1+\xi)}{2(1-\xi)^3}(\log\xi)^3\right\}.$$

Indeed, since by (5.38)

$$\hat{\nu}_{UMVU}^{\hat{\theta}_{ML},X_{(1)}} = X_{(n)}\left[1 + \frac{1}{n\hat{\theta}_{ML}}\left\{\left(\frac{X_{(n)}}{X_{(1)}}\right)^{\hat{\theta}_{ML}} - 1\right\}\right],$$

$T_{(1)} = n(X_{(1)} - \gamma)$, $T_{(n)} = n(X_{(n)} - \nu)$ and $\hat{U} = \sqrt{\lambda_2 n}(\hat{\theta}_{ML} - \theta)$ with $\lambda_2 = \lambda_2(\theta, \gamma, \nu)$, it follows that

$$\begin{aligned}
n(\hat{\nu}_{UMVU}^{\hat{\theta}_{ML},X_{(1)}} - \nu) &= T_{(n)} + \frac{\nu + (T_{(n)}/n)}{\theta + (\hat{U}/\sqrt{\lambda_2 n})}\left\{\left(\frac{\nu + (T_{(n)}/n)}{\gamma + (T_{(1)}/n)}\right)^{\theta + (\hat{U}/\sqrt{\lambda_2 n})} - 1\right\} \\
&= T_{(n)} + \frac{\nu(1-\xi)}{\theta\xi} - \frac{\nu\hat{U}}{\theta^2\xi\sqrt{\lambda_2 n}}(1-\xi+\log\xi) + \frac{\nu}{\xi n}\left(\frac{T_{(n)}}{\nu} - \frac{T_{(1)}}{\gamma}\right) \\
&\quad + \frac{1-\xi}{\theta\xi n}T_{(n)} + \frac{\nu\hat{U}^2}{\theta^3\xi\lambda_2 n}\left\{1-\xi+(\log\xi)\left(\frac{1}{2}\log\xi + 1\right)\right\} \\
&\quad + O_p\left(\frac{1}{n\sqrt{n}}\right),
\end{aligned}$$

hence, by Lemma 3.8.1 in Appendix B1 and Lemma 5.9.1 in Appendix D given later, we have (5.41). Next, we consider the estimator $\hat{\nu}_{UMVU}^{0,X_{(1)}}$ which is treated by Zhang (2013). Since

$$\lim_{\theta\to 0}\frac{1}{\theta}\left\{\left(\frac{X_{(n)}}{\gamma}\right)^{\theta} - 1\right\} = \log\frac{X_{(n)}}{\gamma},$$

it follows from (5.36) that

$$\hat{\nu}_{UMVU}^{0,X_{(1)}} = X_{(n)}\left(1 + \frac{1}{n}\log\frac{X_{(n)}}{X_{(1)}}\right).$$

Then

$$E_{0,\gamma,\nu}\left[n(\hat{\nu}_{UMVU}^{0,X_{(1)}} - \nu)\right] = -\frac{\nu}{n}\log\frac{\gamma}{\nu} + O\left(\frac{1}{n^2}\right), \tag{5.42}$$

which is not asymptotically unbiased up to the order $o(1/n)$. Indeed, since $T_{(1)} = n(X_{(1)} - \gamma)$ and $T_{(n)} = n(X_{(n)} - \nu)$, we have

$$\begin{aligned} n(\hat{\nu}_{UMVU}^{0,X_{(1)}} - \nu) &= T_{(n)} + \left(\nu + \frac{T_{(n)}}{n}\right)\log\frac{\nu + (T_{(n)}/n)}{\gamma + (T_{(1)}/n)} \\ &= \nu\log\frac{\nu}{\gamma} + T_{(n)} + \frac{1}{n}\left\{\left(1 + \log\frac{\nu}{\gamma}\right)T_{(n)} - \frac{\nu}{\gamma}T_{(1)}\right\} + O_p\left(\frac{1}{n^2}\right). \end{aligned} \tag{5.43}$$

Since, by Lemma 3.8.1 in Appendix B1,

$$E_{0,\gamma,\nu}(T_{(1)}) = \gamma\log\frac{\gamma}{\nu} + O\left(\frac{1}{n}\right),$$

$$E_{0,\gamma,\nu}(T_{(n)}) = -\nu\log\frac{\nu}{\gamma} + \frac{\nu}{n}\left(\log\frac{\nu}{\gamma}\right)\left(1 + \log\frac{\nu}{\gamma}\right) + O\left(\frac{1}{n^2}\right),$$

it follows from (5.43) that (5.42) holds. Therefore, it seems to be inappropriate to compare the plug-in estimators $\hat{\nu}_{UMVU}^{\hat{\theta}_{ML},X_{(1)}}$ and $\hat{\nu}_{UMVU}^{0,X_{(1)}}$ with the UMVU estimator $\hat{\nu}_{UMVU}^{\theta,\gamma}$ up to the higher order. Here, we use the bias-adjusted MLE $\hat{\nu}_{ML^*} = X_{(n)}^{**}$ instead of $\hat{\nu}_{UMVU}^{\hat{\theta}_{ML},X_{(1)}}$ and $\hat{\nu}_{UMVU}^{0,X_{(1)}}$. Then, it follows from (5.37) and (5.38) that

$$d_n(\hat{\nu}_{ML^*}, \hat{\nu}_{UMVU}^{\theta,\gamma}) = \frac{(1 - \xi + \log\xi)^2}{(1-\xi)^2 - \xi(\log\xi)^2} + o(1) = d(\xi) + o(1) \quad \text{(say)} \tag{5.44}$$

as $n \to \infty$, where $\xi = (\gamma/\nu)^\theta$. Here, note that $0 < \xi < 1$, $\xi \to 1$ as $\theta \to 0$ and $\xi \to 0$ as $\theta \to \infty$. Then, we have from (5.44)

$$\lim_{\theta\to 0} d(\xi) = 3, \quad \lim_{\theta\to\infty} d(\xi) = \infty,$$

which shows that the second-order asymptotic loss of $\hat{\nu}_{ML^*}$ relative to $\hat{\nu}_{UMVU}^{\theta,\gamma}$ is close to 3 for small θ, but it becomes infinite for large θ. As is seen in the above, a similar consideration to Zhang (2013) from the Monte Carlo simulation seems to be theoretically confirmed. It is also noted from Lemma 3.8.1 in Appendix B1 that the second-order asymptotic mean of $T_{(n)} = n(X_{(n)} - \nu)$ is given by

$$E_{\theta,\gamma,\nu}(T_{(n)}) = -\frac{\nu}{\theta}\left\{\left(\frac{\nu}{\gamma}\right)^{\theta} - 1\right\}$$
$$+ \frac{\nu}{\theta^2 n}\left\{\left(\frac{\nu}{\gamma}\right)^{\theta} - 1\right\}\left\{(\theta+1)\left(\frac{\nu}{\gamma}\right)^{\theta} - 1\right\} + O\left(\frac{1}{n^2}\right)$$

where the first term coincides with the result of Zhang (2013). For the UMVU estimator $\hat{\gamma}_{UMVU}^{\theta,\nu}$, a similar discussion to the above could be done.

Example 5.7.4 (**Two-sided truncated beta distribution**) (Continued from Example 3.7.4). Let $c = 0$, $d = 1$, $a(x) = x^{-1}$, and $u(x) = \log x$ for $0 < \gamma \le x \le \nu < 1$ in the density (1.9). In Example 3.7.4, we have $b(\theta, \gamma) = \theta^{-1}\nu^{\theta}(1 - (\gamma/\nu)^{\theta})$ for $\theta \in \Theta = (0, \infty)$ and the formulae of $\lambda_j(\theta, \gamma, \nu)$ $(j = 1, 2)$, $k(\theta, \gamma, \nu)$, and $\tilde{k}(\theta, \gamma, \nu)$. When θ and γ are known, we obtain by (5.4), (5.5), and (5.7)

$$\hat{\gamma}_{ML^*}^{\theta,\gamma} = X_{(n)}^* = X_{(n)} + \frac{X_{(n)}}{\theta n}\left\{1 - \left(\frac{\gamma}{X_{(n)}}\right)^{\theta}\right\},$$

$$T_{(n)}^* = T_{(n)} + \frac{\nu}{\theta}\left\{1 - \left(\frac{\gamma}{\nu}\right)^{\theta}\right\} - \frac{1}{\theta}\left\{(\theta - 1)\left(1 - \left(\frac{\gamma}{\nu}\right)^{\theta}\right) - \nu \log \nu\right\} T_{(n)}$$
$$+ O_p\left(\frac{1}{n^2}\right),$$

$$V_{\nu}(\tilde{k}T_{(n)}^*) = 1 + \frac{2}{\theta n}\left\{(\theta - 1)\left(1 - \left(\frac{\gamma}{\nu}\right)^{\theta}\right) - \nu \log \nu\right\} + O\left(\frac{1}{n^2}\right).$$

If $\theta = 1$, then the density (1.9) is uniform over the interval $[\gamma, \nu]$. Further, if γ is known, then

$$\hat{\nu}_{ML^*}^{1,\gamma} = \frac{n+1}{n} X_{(n)} - \frac{\gamma}{n},$$

which coincides with the UMVU estimation of ν. When θ and γ are unknown, in a similar way to the above examples, we have the formulae of $\hat{\nu}_{ML^*} (= X_{(n)}^{**})$, $T_{(n)}^{**}$, and $V_{\theta,\gamma,\nu}(\tilde{k}T_{(n)}^{**})$ from (5.25), (5.26), and (5.28). Further, it follows from (5.30) that the second-order asymptotic loss of $\hat{\nu}_{ML^*}$ relative to $\hat{\nu}_{ML^*}^{\theta,\gamma}$ is given by

$$d_n(\hat{\nu}_{ML^*}, \hat{\nu}_{ML^*}^{\theta,\gamma}) = \frac{(1 - \xi + \xi \log \xi)^2}{(1 - \xi)^2 - \xi(\log \xi)^2} + o(1)$$

as $n \to \infty$, where $\xi = (\gamma/\nu)^{\theta}$ (see also Example 3.7.4).

Example 5.7.5 (**Two-sided truncated Erlang distribution**) (Continued from Example 3.7.5). Let $c = 0$, $d = \infty, a(x) = x^{j-1}$, and $u(x) = -x$ for $0 < \gamma \le x \le \nu < \infty$ in the density (1.9), where $j = 1, 2, \dots$. In Example 3.7.5, we have for each $j = 1, 2, \dots, b_j(\theta, \gamma, \nu) = \int_\gamma^\nu x^{j-1}e^{-\theta x}dx$ for $\theta \in \Theta = (0, \infty)$ and the formulae of $\lambda_{ji}(\theta, \gamma, \nu)$ $(i = 1, 2)$, $k_j(\theta, \gamma, \nu)$ and $\tilde{k}_j(\theta, \gamma, \nu)$. Let j be arbitrarily fixed in $\{1, 2, \dots\}$. Since $\tilde{k}_j(\theta, \gamma, \nu) = \nu^{j-1}e^{-\theta\nu}/b_j(\theta, \gamma, \nu)$, we obtain

$$\frac{\partial}{\partial \nu} \log \tilde{k}_j = \frac{j-1}{\nu} - \theta - \frac{1}{b_j}\nu^{j-1}e^{-\theta\nu},$$

where $\tilde{k}_j = \tilde{k}_j(\theta, \gamma, \nu)$ and $b_j = b_j(\theta, \gamma, \nu)$. When θ and γ are known, we have by (5.4), (5.5), and (5.7)

$$\hat{\nu}_{ML^*}^{\theta,\gamma} = X_{(n)}^* = X_{(n)} + \frac{e^{\theta X_{(n)}}b_j(\theta, \gamma, X_{(n)})}{X_{(n)}^{j-1}n},$$

$$T_{(n)}^* = T_{(n)} + \frac{b_j e^{\theta\nu}}{\nu^{j-1}} - \frac{1}{n}\left\{\frac{b_j e^{\theta\nu}}{\nu^{j-1}}\left(\frac{j-1}{\nu} - \theta\right) - 1\right\} T_{(n)} + O_p\left(\frac{1}{n^2}\right),$$

$$V_\nu(\tilde{k}_j T_{(n)}^*) = 1 + \frac{2}{n}\left\{\frac{b_j e^{\theta\nu}}{\nu^{j-1}}\left(\frac{j-1}{\nu} - \theta\right) - 1\right\} + O\left(\frac{1}{n^2}\right).$$

When θ and γ are unknown, in a similar way to the above, we have the formulae of $\hat{\nu}_{ML^*}(= X_{(n)}^{**})$, $T_{(n)}^{**}$, and $V_{\theta,\gamma,\nu}(\tilde{k}_j T_{(n)}^{**})$ from (5.25), (5.26), and (5.28). Further, it follows from (5.30) that the second-order asymptotic loss of $\hat{\nu}_{ML^*}$ relative to $\hat{\nu}_{ML^*}^{\theta,\gamma}$ is given by

$$d_n(\hat{\nu}_{ML^*}, \hat{\nu}_{ML^*}^{\theta,\gamma}) = \left(\frac{b_{j+1}}{b_j} - \nu\right)^2 \Bigg/ \left\{\frac{b_{j+2}}{b_j} - \left(\frac{b_{j+1}}{b_j}\right)^2\right\} + o(1)$$

as $n \to \infty$.

A similar example to the above in the two-sided truncated lognormal case in Example 3.7.6 is reduced to the two-sided truncated normal one in Example 5.7.2.

5.8 Concluding Remarks

In Chap. 3, for a tTEF of distributions with two truncation parameters γ, ν and a natural parameter θ including the upper-truncated Pareto distribution, we discussed the estimation problem on θ together with a bias-adjustment when γ and ν are known or unknown nuisance parameters. In this chapter, exchanging the situation, we considered the estimation on ν when θ and γ were known or unknown nuisance parameters. Indeed, we obtained the bias-adjusted MLE $\hat{\nu}_{ML^*}^{\theta,\gamma}$ of ν when θ and γ were known, the bias-adjusted MLE $\hat{\nu}^{\gamma}_{ML^*}$ of ν when θ was unknown and γ was known, and the bias-adjusted MLE $\hat{\nu}_{ML^*}$ of ν when θ and γ were unknown, and derived their stochastic expansions and second-order asymptotic variances. Further we got the second-order asymptotic losses of $\hat{\nu}^{\gamma}_{ML^*}$ and $\hat{\nu}_{ML^*}$ relative to $\hat{\nu}^{\theta,\gamma}_{ML^*}$. On the bias-adjusted MLE $\hat{\nu}^{\gamma}_{ML^*}$ of ν when θ was unknown and γ was known, the situation was reduced to the oTEF of distributions with a lower truncation parameter and a natural parameter θ which was discussed in Chap. 4. In this case, we needed a transformation $Y = -X$ which was described in Remark 4.5.3 and was also carried out in Sect. 5.4 and in a similar way to the section we obtained a similar result. As an example, we treated the upper-truncated Pareto case, where the results based on the Monte Carlo simulation by Zhang (2013) were theoretically confirmed in this chapter.

5.9 Appendix D

Before proving Theorems 5.3.1 and 5.4.2, we prepare a lemma.

Lemma 5.9.1 *Let $\hat{U} = \sqrt{\lambda_2 n}(\hat{\theta}_{ML} - \theta)$. Then, the asymptotic expectation of $\hat{U}$ and $\hat{U}^2$ is given by*

$$E_{\theta,\gamma,\nu}(\hat{U}) = -\frac{1}{\sqrt{\lambda_2 n}}\left\{\frac{1}{k}\left(\frac{\partial\lambda_1}{\partial\gamma}\right) - \frac{1}{\tilde{k}}\left(\frac{\partial\lambda_1}{\partial\nu}\right) + \frac{\lambda_3}{2\lambda_2}\right\} + O\left(\frac{1}{n\sqrt{n}}\right), \tag{5.45}$$

$$E_{\theta,\gamma,\nu}(\hat{U}^2) = 1 + O\left(\frac{1}{n}\right), \tag{5.46}$$

where $\lambda_i = \lambda_i(\theta, \gamma, \nu)$ $(j = 1, 2, 3)$, $k = k(\theta, \gamma, \nu)$, and $\tilde{k} = \tilde{k}(\theta, \gamma, \nu)$.

The Eqs. (5.45) and (5.46) are obtained as (3.28) and (3.31), respectively.

The proof of Theorem 5.3.1 By the Taylor expansion, we have

$$\frac{1}{\hat{\tilde{k}}_{\theta,\gamma}} = \frac{1}{\tilde{k}(\theta, \gamma, X_{(n)})} = \frac{1}{\tilde{k}(\theta, \gamma, \nu)}\left[1 - \left\{\frac{\partial}{\partial\nu}\log\tilde{k}(\theta, \gamma, \nu)\right\}\frac{T_{(n)}}{n} + O_p\left(\frac{1}{n^2}\right)\right]. \tag{5.47}$$

Substituting (5.47) into (5.4), we obtain from (5.3)

$$X^*_{(n)} = X_{(n)} + \frac{1}{\tilde{k}n} + \frac{1}{n^2}\tilde{A} - \frac{1}{\tilde{k}n^2}\left(\frac{\partial}{\partial\nu}\log\tilde{k}\right)\left(T_{(n)} + \frac{1}{\tilde{k}}\right) + O_p\left(\frac{1}{n^3}\right),$$

where $\tilde{k} = \tilde{k}(\theta, \gamma, \nu)$ and $\tilde{A} = \tilde{A}(\theta, \gamma, \nu)$, hence by (5.3)

$$T^*_{(n)} = n(X^*_{(n)} - \nu) = T_{(n)} + \frac{1}{\tilde{k}} + \frac{1}{n}\tilde{A} - \frac{1}{\tilde{k}n}\left(\frac{\partial}{\partial\nu}\log\tilde{k}\right)\left(T_{(n)} + \frac{1}{\tilde{k}}\right) + O_p\left(\frac{1}{n^2}\right)$$
$$= T_{(n)} + \frac{1}{\tilde{k}} - \frac{1}{\tilde{k}n}\left(\frac{\partial}{\partial\nu}\log\tilde{k}\right)T_{(n)} + O_p\left(\frac{1}{n^2}\right).$$

Thus, we get (5.5). From (3.17), (3.19) in Appendix B1, and (5.3), it follows that the second-order asymptotic mean and variance of $T^*_{(n)}$ are

$$E_\nu(T^*_{(n)}) = O\left(\frac{1}{n^2}\right), \quad V_\nu(T^*_{(n)}) = \frac{1}{\tilde{k}^2} + \frac{2}{\tilde{k}^3 n}\left(\frac{\partial}{\partial\nu}\log\tilde{k}\right) + O\left(\frac{1}{n^2}\right),$$

hence, we get (5.7). Thus, we complete the proof.

The proof of Theorem 5.4.1 Since $\delta = -\nu$, $\eta = -\gamma$, and $X_{(n)} = -Y_{(1)}$, it follows from (5.14) and (5.15) that $\hat{k}_{0\eta} = \hat{\tilde{k}}_\gamma$, $\hat{\lambda}^0_{\eta j} = \hat{\lambda}_{\gamma j}$ $(j = 2, 3)$, $\partial^j\hat{k}_{0\eta}/\partial\theta^j = \partial^j\hat{\tilde{k}}_\gamma/\partial\theta^j$ $(j = 1, 2)$, and $\partial\hat{\lambda}^0_{\eta 1}/\partial\delta = -(\partial\hat{\lambda}_{\gamma 1}/\partial\nu)$, hence, letting $X^\dagger_{(n)} = -Y'_{(1)}$, we have from (5.16)

$$X^\dagger_{(n)} = X_{(n)} + \frac{1}{\hat{\tilde{k}}_\gamma n} + \frac{1}{\hat{\tilde{k}}^2_\gamma\hat{\lambda}_{\gamma 2}n^2}\left(\frac{\partial\hat{\tilde{k}}_\gamma}{\partial\theta}\right)\left\{\frac{1}{\hat{\tilde{k}}_\gamma}\left(\frac{\partial\hat{\lambda}_{\gamma 1}}{\partial\nu}\right) - \frac{\hat{\lambda}_{\gamma 3}}{2\hat{\lambda}_{\gamma 2}}\right\}$$
$$+ \frac{1}{2\hat{\tilde{k}}^2_\gamma\hat{\lambda}_{\gamma 2}n^2}\left\{\frac{\partial^2\hat{\tilde{k}}_\gamma}{\partial\theta^2} - \frac{2}{\hat{\tilde{k}}_\gamma}\left(\frac{\partial\hat{\tilde{k}}_\gamma}{\partial\theta}\right)^2\right\},$$

which coincides with (5.20), where

$$\hat{\tilde{k}}_\gamma = \tilde{k}_\gamma\left(\hat{\theta}_{ML}, X_{(n)}\right) = \tilde{k}\left(\hat{\theta}_{ML}, \gamma, X_{(n)}\right),$$
$$\partial^j\hat{\tilde{k}}_\gamma/\partial\theta^j = \left(\partial^j/\partial\theta^j\right)\tilde{k}_\gamma\left(\hat{\theta}_{ML}, X_{(n)}\right) = \left(\partial^j/\partial\theta^j\right)\tilde{k}\left(\hat{\theta}_{ML}, \gamma, X_{(n)}\right) \quad (j = 1, 2),$$
$$\hat{\lambda}_{\gamma j} = \lambda_{\gamma j}\left(\hat{\theta}_{ML}, X_{(n)}\right) = \lambda_j\left(\hat{\theta}_{ML}, \gamma, X_{(n)}\right) \quad (j = 2, 3),$$
$$\partial\hat{\lambda}_{\gamma 1}/\partial\nu = (\partial/\partial\nu)\,\lambda_{\gamma 1}\left(\hat{\theta}_{ML}, X_{(n)}\right) = (\partial/\partial\nu)\,\lambda_1\left(\hat{\theta}_{ML}, \gamma, X_{(n)}\right),$$

and also from (5.17)

$$T^\dagger_{(n)} = T_{(n)} + \frac{1}{\tilde{k}_\gamma} - \frac{1}{\tilde{k}^2_\gamma\sqrt{\lambda_{\gamma 2}n}}\left(\frac{\partial\tilde{k}_\gamma}{\partial\theta}\right)\left[\hat{U}_\gamma + \frac{1}{\sqrt{\lambda_{\gamma 2}n}}\left\{-\frac{1}{\tilde{k}_\gamma}\left(\frac{\partial\lambda_{\gamma 1}}{\partial\nu}\right) + \frac{\lambda_{\gamma 3}}{2\lambda_{\gamma 2}}\right\}\right]$$
$$- \frac{1}{\tilde{k}^2_\gamma n}\left(\frac{\partial\tilde{k}_\gamma}{\partial\nu}\right)T_{(n)} - \frac{1}{2\tilde{k}^2_\gamma\lambda_{\gamma 2}n}\left\{\frac{\partial^2\tilde{k}_\gamma}{\partial\theta^2} - \frac{2}{\tilde{k}_\gamma}\left(\frac{\partial\tilde{k}_\gamma}{\partial\theta}\right)^2\right\}\left(\hat{U}^2_\gamma - 1\right)$$
$$+ O_p\left(\frac{1}{n\sqrt{n}}\right),$$

hence, (5.21) holds. Since $T'_{(1)} = n(Y'_{(1)} - \delta) = -n(X_{(n)} - \nu) = -T_{(n)}$, $(\partial/\partial\delta)\log k_{0\eta} = -(\partial/\partial\nu)\log \tilde{k}_\gamma$ by (5.14) and $u_0(\delta) = u(\nu)$, it follows from (5.15), (5.18), and (5.19) that (5.22) and (5.23) hold. Thus, we complete the proof.

The proof of Theorem 5.5.1 Putting $T_{(1)} := n(X_{(1)} - \gamma)$, $T_{(n)} := n(X_{(n)} - \nu)$, and

$$Z_1 := \frac{1}{\sqrt{\lambda_2(\theta,\gamma,\nu)n}}\sum_{i=1}^n \{u(X_i) - \lambda_1(\theta,\gamma,\nu)\},$$

we have from (3.7) in Theorem 3.4.1

$$\hat{U} = Z_1 - \frac{\lambda_3}{2\lambda_2^{3/2}\sqrt{n}}Z_1^2 - \frac{1}{\sqrt{\lambda_2 n}}\left\{\left(\frac{\partial\lambda_1}{\partial\gamma}\right)T_{(1)} + \left(\frac{\partial\lambda_1}{\partial\nu}\right)T_{(n)}\right\} + O_p\left(\frac{1}{n}\right). \tag{5.48}$$

By the Taylor expansion, we have

$$\begin{aligned}
&\tilde{k}(\hat{\theta}_{ML}, X_{(1)}, X_{(n)}) \\
&= \tilde{k} + \frac{1}{\sqrt{\lambda_2 n}}\left(\frac{\partial\tilde{k}}{\partial\theta}\right)\hat{U} + \frac{1}{n}\left(\frac{\partial\tilde{k}}{\partial\gamma}\right)T_{(1)} + \frac{1}{n}\left(\frac{\partial\tilde{k}}{\partial\nu}\right)T_{(n)} + \frac{1}{2\lambda_2 n}\left(\frac{\partial^2\tilde{k}}{\partial\theta^2}\right)\hat{U}^2 \\
&+ O_p\left(\frac{1}{n\sqrt{n}}\right),
\end{aligned}$$

where $\tilde{k} = \tilde{k}(\theta,\gamma,\nu)$, $\partial\tilde{k}/\partial\theta = (\partial/\partial\theta)\tilde{k}(\theta,\gamma,\nu)$, $\partial\tilde{k}/\partial\gamma = (\partial/\partial\gamma)\tilde{k}(\theta,\gamma,\nu)$, $\partial\tilde{k}/\partial\nu = (\partial/\partial\nu)\tilde{k}(\theta,\gamma,\nu)$, and $\partial^2\tilde{k}/\partial\theta^2 = (\partial^2/\partial\theta^2)\tilde{k}(\theta,\gamma,\nu)$. Since

$$\begin{aligned}
\frac{1}{\hat{\tilde{k}}} &= \frac{1}{\tilde{k}(\hat{\theta}_{ML}, X_{(1)}, X_{(n)})} \\
&= \frac{1}{\tilde{k}} - \frac{1}{\tilde{k}^2\sqrt{\lambda_2 n}}\left(\frac{\partial\tilde{k}}{\partial\theta}\right)\hat{U} - \frac{1}{\tilde{k}^2 n}\left(\frac{\partial\tilde{k}}{\partial\gamma}\right)T_{(1)} - \frac{1}{\tilde{k}^2 n}\left(\frac{\partial\tilde{k}}{\partial\nu}\right)T_{(n)} - \frac{1}{2\tilde{k}^2\lambda_2 n}\left(\frac{\partial^2\tilde{k}}{\partial\theta^2}\right)\hat{U}^2 \\
&+ \frac{1}{\tilde{k}^3\lambda_2 n}\left(\frac{\partial\tilde{k}}{\partial\theta}\right)^2\hat{U}^2 + O_p\left(\frac{1}{n\sqrt{n}}\right),
\end{aligned}$$

$$\hat{k} = k(\hat{\theta}_{ML}, X_{(1)}, X_{(n)}) = k(\theta,\gamma,\nu) + O_p\left(\frac{1}{\sqrt{n}}\right) = k + O_p\left(\frac{1}{\sqrt{n}}\right),$$

$$\hat{\lambda}_j = \lambda_j(\hat{\theta}_{ML}, X_{(1)}, X_{(n)}) = \lambda_j(\theta,\gamma,\nu) + O_p\left(\frac{1}{\sqrt{n}}\right) = \lambda_j + O_p\left(\frac{1}{\sqrt{n}}\right) \quad (j = 1, 2, 3),$$

$$\frac{\partial\hat{\lambda}_1}{\partial\gamma} = \frac{\partial\lambda_1}{\partial\gamma}\left(\hat{\theta}_{ML}, X_{(1)}, X_{(n)}\right) = \frac{\partial\lambda_1}{\partial\gamma}(\theta,\gamma,\nu) + O_p\left(\frac{1}{\sqrt{n}}\right) = \frac{\partial\lambda_1}{\partial\gamma} + O_p\left(\frac{1}{\sqrt{n}}\right),$$

$$\frac{\partial\hat{\lambda}_1}{\partial\nu} = \frac{\partial\lambda_1}{\partial\nu}(\hat{\theta}_{ML}, X_{(1)}, X_{(n)}) = \frac{\partial\lambda_1}{\partial\nu}(\theta,\gamma,\nu) + O_p\left(\frac{1}{\sqrt{n}}\right) = \frac{\partial\lambda_1}{\partial\nu} + O_p\left(\frac{1}{\sqrt{n}}\right),$$

$$\frac{\partial \hat{\tilde{k}}}{\partial \gamma} = \frac{\partial \tilde{k}}{\partial \gamma}(\hat{\theta}_{ML}, X_{(1)}, X_{(n)}) = \frac{\partial \tilde{k}}{\partial \gamma}(\theta, \gamma, \nu) + O_p\left(\frac{1}{\sqrt{n}}\right) = \frac{\partial \tilde{k}}{\partial \gamma} + O_p\left(\frac{1}{\sqrt{n}}\right),$$

$$\frac{\partial^j \hat{\tilde{k}}}{\partial \theta^j} = \frac{\partial^j \tilde{k}}{\partial \theta^j}(\hat{\theta}_{ML}, X_{(1)}, X_{(n)}) = \frac{\partial^j \tilde{k}}{\partial \theta^j}(\theta, \gamma, \nu) + O_p\left(\frac{1}{\sqrt{n}}\right) = \frac{\partial^j \tilde{k}}{\partial \theta^j} + O_p\left(\frac{1}{\sqrt{n}}\right)$$
$$(j = 1, 2),$$

substituting them into (5.25), we have

$$\begin{aligned}T_{(n)}^{**} &= n(X_{(n)}^{**} - \nu)\\
&= T_{(n)} + \frac{1}{\tilde{k}} - \frac{1}{\tilde{k}^2\sqrt{\lambda_2 n}}\left(\frac{\partial \tilde{k}}{\partial \theta}\right)\left[\hat{U} + \frac{1}{\sqrt{\lambda_2 n}}\left\{\frac{1}{k}\left(\frac{\partial \lambda_1}{\partial \gamma}\right) - \frac{1}{\tilde{k}}\left(\frac{\partial \lambda_1}{\partial \nu}\right) + \frac{\lambda_3}{2\lambda_2}\right\}\right]\\
&\quad - \frac{1}{\tilde{k}n}\left(\frac{\partial}{\partial \gamma}\log \tilde{k}\right)\left(T_{(1)} - \frac{1}{k}\right) - \frac{1}{\tilde{k}n}\left(\frac{\partial}{\partial \nu}\log \tilde{k}\right)T_{(n)}\\
&\quad - \frac{1}{2\tilde{k}^2\lambda_2 n}\left\{\frac{\partial^2 \tilde{k}}{\partial \theta^2} - \frac{2}{\tilde{k}}\left(\frac{\partial \tilde{k}}{\partial \theta}\right)^2\right\}(U^2 - 1) + O_p\left(\frac{1}{n\sqrt{n}}\right),\end{aligned} \tag{5.49}$$

which shows that (5.26) holds. From (5.26) and Lemmas 3.9.1 and 5.9.1, we obtain (5.27). Since $T_{(1)}$ and $T_{(n)}$ are asymptotically independent, it follows from (3.16) and (3.17) that

$$E_{\theta,\gamma,\nu}\left[T_{(1)}T_{(n)}\right] = -\frac{1}{k\tilde{k}} + O\left(\frac{1}{n}\right). \tag{5.50}$$

By (5.48), (5.50), and Lemma 5.9.1, we obtain

$$E_{\theta,\gamma,\nu}\left[\left(T_{(n)} + \frac{1}{\tilde{k}}\right)\hat{U}\right] = O\left(\frac{1}{n}\right), \tag{5.51}$$

since $\partial\lambda_1/\partial\nu = \tilde{k}(u(\nu) - \lambda_1)$. Since $E_{\theta,\gamma,\nu}(Z_1^2|T_{(n)}) = 1 + O_p(1/n)$ by (2.59), it follows from (3.17) and (5.48) that

$$E_{\theta,\gamma,\nu}\left[\left(T_{(n)} + \frac{1}{\tilde{k}}\right)(\hat{U}^2 - 1)\right] = O\left(\frac{1}{n\sqrt{n}}\right). \tag{5.52}$$

By (5.49)–(5.52), Lemmas 3.9.1, and 5.9.1 and (5.3), we have

$$\begin{aligned}&V_{\theta,\gamma,\nu}(T_{(n)}^{**})\\
&= \left\{1 - \frac{2}{\tilde{k}n}\left(\frac{\partial}{\partial \nu}\log \tilde{k}\right)\right\}E_{\theta,\gamma,\nu}\left[\left(T_{(n)} + \frac{1}{\tilde{k}}\right)^2\right] + \frac{1}{\tilde{k}^4\lambda_2 n}\left(\frac{\partial \tilde{k}}{\partial \theta}\right)^2 E_{\theta,\gamma,\nu}(\hat{U}^2)\end{aligned}$$

$$+ O\left(\frac{1}{n\sqrt{n}}\right)$$

$$= \frac{1}{\tilde{k}^2} + \frac{2}{\tilde{k}^3 n}\left(\frac{\partial}{\partial \nu}\log \tilde{k}\right) + \frac{1}{\tilde{k}^2 \lambda_2 n}\{u(\nu) - \lambda_1\}^2 + O\left(\frac{1}{n\sqrt{n}}\right),$$

since

$$\frac{\partial}{\partial \theta}\log \tilde{k} = u(\nu) - \frac{\partial}{\partial \theta}\log b(\theta, \gamma, \nu) = u(\nu) - \lambda_1,$$

which shows that (5.28) holds. Thus, we complete the proof.

References

Akahira, M., & Ohyauchi, N. (2016). Second order asymptotic loss of the MLE of a truncation parameter for a two-sided truncated exponential family of distributions. *Journal of the Japan Statistical Society*, *46*, 27–50.

Zhang, J. (2013). Reducing bias of the maximum-likelihood estimation for the truncated Pareto distribution. *Statistics*, *47*, 792–799.

Chapter 6
Bayesian Estimation of a Truncation Parameter for a One-Sided TEF

For a one-sided truncated exponential family (oTEF) of distributions with a truncation parameter γ and a natural parameter θ as a nuisance parameter, the asymptotic behavior of the Bayes estimator of γ is discussed.

6.1 Introduction

In this chapter, following mostly the paper by Akahira (2016), the estimation problem on γ for a oTEF of distributions is considered from the Bayesian viewpoint. Under a quadratic loss and a smooth prior on γ, the Bayes estimator of γ is well known to be expressed as a form of the posterior mean. In Sect. 6.3, when θ is known, the stochastic expansion of the Bayes estimator $\hat{\gamma}_{B,\theta}$ of γ is derived, and the second order asymptotic mean and asymptotic variance of $\hat{\gamma}_{B,\theta}$ are given. In Sect. 6.4, when θ is unknown, the stochastic expansion of the Bayes estimator $\hat{\gamma}_{B,\hat{\theta}_{ML}}$ plugging the MLE $\hat{\theta}_{ML}$ in θ of $\hat{\gamma}_{B,\theta}$ is derived, and the second-order asymptotic mean and asymptotic variance of $\hat{\gamma}_{B,\hat{\theta}_{ML}}$ are given. In Sect. 6.5, several examples for a lower-truncated exponential, a lower-truncated normal, Pareto, a lower-truncated beta, and a lower-truncated Erlang distributions are given. In Appendix E, the proofs of Theorems 6.3.1 and 6.4.1 are given.

6.2 Formulation and Assumptions

Suppose that $X_1, X_2, \ldots, X_n, \ldots$ is a sequence of i.i.d. random variables according to $P_{\theta,\gamma}$, having the density (1.7), which belongs to a oTEF of distributions. Let $\pi(\gamma)$ be a prior density with respect to the Lebesgue measure over the open interval

M. Akahira, *Statistical Estimation for Truncated Exponential Families*,
JSS Research Series in Statistics, DOI 10.1007/978-981-10-5296-5_6

(c, d), and $L(\hat{\gamma}, \gamma)$ the quadratic loss $(\hat{\gamma} - \gamma)^2$ of any estimator $\hat{\gamma} = \hat{\gamma}(\boldsymbol{X})$ based on $\boldsymbol{X} := (X_1, \ldots, X_n)$. Suppose that θ is known. Then, it is easily seen that the Bayes estimator of γ w.r.t. L and π is given by

$$\hat{\gamma}_{B,\theta}(\boldsymbol{X}) := \int_c^{X_{(1)}} \frac{t\pi(t)}{b^n(\theta, t)} dt \bigg/ \int_c^{X_{(1)}} \frac{\pi(t)}{b^n(\theta, t)} dt, \tag{6.1}$$

where $X_{(1)} := \min_{1 \le i \le n} X_i$. In what follows, we always assume that $a(\cdot)$ and $u(\cdot)$ are functions of class C^3 and $\pi(\cdot)$ is a function of class C^2 on the interval (c, d), where C^k is the class of all k times continuously differentiable functions for any positive integer k.

6.3 Bayes Estimator $\hat{\gamma}_{B,\theta}$ of γ When θ is Known

Letting $u = n(t - \gamma)$, we have from (6.1)

$$\hat{\gamma}_{B,\theta}(\boldsymbol{X}) = \gamma + \frac{1}{n}\left(\int_{\tau_n}^{T_{(1)}} \frac{u\pi\left(\gamma + (u/n)\right)}{b^n\left(\theta, \gamma + (u/n)\right)} du \bigg/ \int_{\tau_n}^{T_{(1)}} \frac{\pi\left(\gamma + (u/n)\right)}{b^n\left(\theta, \gamma + (u/n)\right)} du \right), \tag{6.2}$$

where $\tau_n := n(c - \gamma)$ and $T_{(1)} := n(X_{(1)} - \gamma)$. Let

$$b_{(j)}(\theta, \gamma) := \frac{\partial^j}{\partial \gamma^j} \log b(\theta, \gamma) \qquad (j = 1, 2, \ldots), \tag{6.3}$$

$$\pi_{(j)}(\gamma) := \frac{\partial^j}{\partial \gamma^j} \log \pi(\gamma) \qquad (j = 1, 2, \ldots). \tag{6.4}$$

It is noted from (1.5) that

$$k(\theta, \gamma) := \frac{a(\gamma)e^{\theta u(\gamma)}}{b(\theta, \gamma)} = -b_{(1)}(\theta, \gamma). \tag{6.5}$$

Then, we have the following.

Theorem 6.3.1 *For a oTEF $\mathscr{P}_o$ of distributions having densities of the form (1.7) with a truncation parameter γ and a natural parameter θ, let $\hat{\gamma}_{B,\theta}$ be the Bayes estimator (6.1) of γ w.r.t. the loss L and the prior density π, when θ is known. Then, the stochastic expansion of $T_{B,\theta} := n(\hat{\gamma}_{B,\theta} - \gamma)$ is given by*

$$T_{B,\theta} = T_{(1)} - \frac{1}{k} + \frac{1}{kn}\left(\frac{\partial}{\partial \gamma} \log k\right) T_{(1)} - \frac{1}{k^2 n}\left\{2\left(\frac{\partial}{\partial \gamma} \log k\right) - \pi_{(1)}\right\} + O_p\left(\frac{1}{n^2}\right), \tag{6.6}$$

and the second-order asymptotic mean and asymptotic variance of $kT_{B,\theta}$ are given by

$$E_\gamma(kT_{B,\theta}) = -\frac{1}{kn}\left\{2\left(\frac{\partial}{\partial\gamma}\log k\right) - \pi_{(1)}\right\} + O\left(\frac{1}{n^2}\right), \tag{6.7}$$

$$V_\gamma(kT_{B,\theta}) = 1 - \frac{2}{kn}\left(\frac{\partial}{\partial\gamma}\log k\right) + O\left(\frac{1}{n^2}\right), \tag{6.8}$$

respectively, where $k = k(\theta, \gamma)$ and $\pi_{(1)} = \pi_{(1)}(\gamma)$.

Remark 6.3.1 The second-order asymptotic variance of $kT_{B,\theta}$ is free of the prior density π. We see that the part except the term

$$-\frac{1}{k^2 n}\left\{2\left(\frac{\partial}{\partial\gamma}\log k\right) - \pi_{(1)}\right\} \tag{6.9}$$

in the right-hand side of (6.6) coincides with the stochastic expansion (4.6) of $T_{(1)}^* := n(X_{(1)}^* - \gamma)$ up to the order n^{-1}, where $X_{(1)}^* := X_{(1)} - (\hat{k}_\theta n)^{-1}$ is the bias-adjusted MLE $\hat{\gamma}_{ML^*}^\theta$, with $\hat{k}_\theta = k(\theta, X_{(1)})$ (see (4.5)). Then, we have

$$k(T_{B,\theta} - T_{(1)}^*) = -\frac{1}{kn}\left\{2\left(\frac{\partial}{\partial\gamma}\log k\right) - \pi_{(1)}\right\} + O_p\left(\frac{1}{n^2}\right),$$

which affects only the difference between their second-order asymptotic means. That is, $\hat{\gamma}_{B,\theta}$ is second order asymptotically equivalent to the bias-adjusted MLE $\hat{\gamma}_{ML^*}^\theta = X_{(1)}^*$, hence by (4.8) and (6.8)

$$V_\gamma(kT_{B,\theta}) - V_\gamma(kT_{(1)}^*) = O\left(\frac{1}{n^2}\right).$$

6.4 Bayes Estimator $\hat{\gamma}_{B,\hat{\theta}_{ML}}$ of γ When θ is Unknown

Let $\hat{\gamma}_{ML}$ and $\hat{\theta}_{ML}$ be the MLEs of γ and θ, respectively. From (4.9) it is seen that $\hat{\gamma}_{ML} = X_{(1)}$ and $L(X_{(1)}, \hat{\theta}_{ML}; \boldsymbol{X}) = \sup_{\theta\in\Theta} L(X_{(1)}, \theta; \boldsymbol{X})$, hence $\hat{\theta}_{ML}$ satisfies the likelihood equation (4.10). Denote by $\lambda_j(\theta, \gamma)$ the j-th cumulant (4.1) corresponding to (1.7) for $j = 1, 2, \ldots$. Let $\lambda_2 = \lambda_2(\theta, \gamma)$ and $\hat{U} = \sqrt{\lambda_2 n}(\hat{\theta}_{ML} - \theta)$. When θ is unknown, using the MLE $\hat{\theta}_{ML}$ of θ we consider the Bayes estimator plugging $\hat{\theta}_{ML}$ in θ of $\hat{\gamma}_{B,\theta}$, i.e.,

$$\hat{\gamma}_{B,\hat{\theta}_{ML}}(\boldsymbol{X}) := \int_c^{X_{(1)}} \frac{t\pi(t)}{b^n(\hat{\theta}_{ML}, t)}dt \Big/ \int_c^{X_{(1)}} \frac{\pi(t)}{b^n(\hat{\theta}_{ML}, t)}dt. \tag{6.10}$$

Then we have the following.

Theorem 6.4.1 *For a oTEF $\mathscr{P}_o$ of distributions having densities of the form (1.7) with a truncation parameter γ and a natural parameter θ, let $\hat{\gamma}_{B,\hat{\theta}_{ML}}$ be the Bayes estimator (6.10) plugging $\hat{\theta}_{ML}$ in θ of $\hat{\gamma}_{B,\theta}$ when θ is unknown. Then the stochastic expansion of $T_{B,\hat{\theta}_{ML}} := n(\hat{\gamma}_{B,\hat{\theta}_{ML}} - \gamma)$ is given by*

$$
\begin{aligned}
T_{B,\hat{\theta}_{ML}} = & T_{(1)} - \frac{1}{k} + \frac{1}{k^2\sqrt{\lambda_2 n}}\left(\frac{\partial k}{\partial \theta}\right)\left\{\hat{U} + \frac{1}{\sqrt{\lambda_2 n}}\left(\frac{1}{k}\left(\frac{\partial \lambda_1}{\partial \gamma}\right) + \frac{\lambda_3}{2\lambda_2}\right)\right\} \\
& + \frac{1}{kn}\left(\frac{\partial}{\partial \gamma}\log k\right)T_{(1)} + \frac{1}{2k^2\lambda_2 n}\left\{\frac{\partial^2 k}{\partial \theta^2} - \frac{2}{k}\left(\frac{\partial k}{\partial \theta}\right)^2\right\}(\hat{U}^2 - 1) + \frac{B}{kn} \\
& + O_p\left(\frac{1}{n\sqrt{n}}\right),
\end{aligned} \tag{6.11}
$$

where

$$
B := -\frac{1}{2\lambda_2}\left\{\frac{\lambda_3}{k\lambda_2}\left(\frac{\partial k}{\partial \theta}\right) - \frac{1}{k}\left(\frac{\partial^2 k}{\partial \theta^2}\right)\right\} - \frac{1}{k}\left\{2\left(\frac{\partial}{\partial \gamma}\log k\right) - \pi_{(1)}\right\},
$$

$k = k(\theta, \gamma)$, $\lambda_j = \lambda_j(\theta, \gamma)$ $(j = 1, 2, 3)$, and the second-order asymptotic mean and asymptotic variance of $kT_{B,\hat{\theta}_{ML}}$ are given by

$$
E_{\theta,\gamma}[kT_{B,\hat{\theta}_{ML}}] = \frac{B}{n} + O\left(\frac{1}{n\sqrt{n}}\right), \tag{6.12}
$$

$$
V_{\theta,\gamma}(kT_{B,\hat{\theta}_{ML}}) = 1 - \frac{2}{kn}\left(\frac{\partial}{\partial \gamma}\log k\right) + \frac{1}{\lambda_2 n}\{u(\gamma) - \lambda_1\}^2 + O\left(\frac{1}{n\sqrt{n}}\right), \tag{6.13}
$$

respectively.

Remark 6.4.1 The second-order asymptotic variance of $kT_{B,\hat{\theta}_{ML}}$ is free of the prior density π. In the stochastic expansion (6.11), the terms involving $\hat{U}$ yield from the use of the MLE $\hat{\theta}_{ML}$ in θ of $\hat{\gamma}_{B,\theta}$. It is also seen from (6.12) and (6.13) that the terms depending on λ_2 and λ_3 in the second-order asymptotic mean and asymptotic variance come from using $\hat{\theta}_{ML}$.

Remark 6.4.2 We see that the part except the term $B/(kn)$ in the right-hand side of (6.11) coincides with the stochastic expansion (4.12) of $T_{(1)}^{**} := n(X_{(1)}^{**} - \gamma)$ up to the order n^{-1}, where

$$
\begin{aligned}
X_{(1)}^{**} := & X_{(1)} - \frac{1}{\hat{k}n} + \frac{1}{\hat{k}^2\hat{\lambda}_2 n^2}\left(\frac{\partial \hat{k}}{\partial \theta}\right)\left\{\frac{1}{\hat{k}}\left(\frac{\partial \hat{\lambda}_1}{\partial \gamma}\right) + \frac{\hat{\lambda}_3}{2\hat{\lambda}_2}\right\} \\
& - \frac{1}{2\hat{k}^2\hat{\lambda}_2 n^2}\left\{\frac{\partial^2 \hat{k}}{\partial \theta^2} - \frac{2}{\hat{k}}\left(\frac{\partial \hat{k}}{\partial \theta}\right)^2\right\}
\end{aligned}
$$

is the bias-adjusted MLE $\hat{\gamma}_{ML^*}$, with $\hat{k} = k(\hat{\theta}_{ML}, X_{(1)})$, $\partial^j \hat{k}/\partial\theta^j = (\partial^j/\partial\theta^j)k(\hat{\theta}_{ML}, X_{(1)})$ $(j = 1, 2)$, $\hat{\lambda}_j = \lambda_j(\hat{\theta}_{ML}, X_{(1)})$ $(j = 2, 3)$ and $\partial\hat{\lambda}_1/\partial\gamma = (\partial/\partial\gamma)\lambda_1(\hat{\theta}_{ML}, X_{(1)})$ (see (4.11)). Then, we have

$$k(T_{B,\hat{\theta}_{ML}} - T^{**}_{(1)}) = \frac{B}{n} + O_p\left(\frac{1}{n\sqrt{n}}\right),$$

which affects only the difference between their second-order asymptotic means. That is, $\hat{\gamma}_{B,\hat{\theta}_{ML}}$ is second-order asymptotic equivalent to the bias-adjusted MLE $\hat{\gamma}_{ML^*} = X^{**}_{(1)}$, hence by (4.14) and (6.13)

$$V_{\theta,\gamma}(kT_{B,\hat{\theta}_{ML}}) - V_{\theta,\gamma}(kT^{**}_{(1)}) = O\left(\frac{1}{n\sqrt{n}}\right).$$

Further, it is easily seen from (6.8) and (6.13), the second-order asymptotic loss of $\hat{\gamma}_{B,\hat{\theta}_{ML}}$ relative to $\hat{\gamma}_{B,\theta}$ is given by

$$d_n(\hat{\gamma}_{B,\hat{\theta}_{ML}}, \hat{\gamma}_{B,\theta}) := n\left\{V_{\theta,\gamma}(kT_{B,\hat{\theta}_{ML}}) - V_\gamma(kT_{B,\theta})\right\} = \frac{1}{\lambda_2}\{u(\gamma) - \lambda_1\}^2 + o(1).$$

6.5 Examples

We consider a lower-truncated exponential, a lower-truncated normal, Pareto, a lower-truncated beta, and a lower-truncated Erlang distributions as in Chaps. 2 and 4.

Example 6.5.1 (**Lower-truncated exponential distribution**) (Continued from Examples 2.7.1 and 4.6.1). Let $c = -\infty$, $d = \infty$, $a(x) \equiv 1$, and $u(x) = -x$ for $-\infty < \gamma \le x < \infty$ in the density (1.7). Since $b(\theta, \gamma) = \theta^{-1}e^{-\theta\gamma}$ for $\theta \in \Theta = (0, \infty)$, it follows from (4.1), (6.3), and (6.5) that $b_{(1)}(\theta, \gamma) = -\theta$, $b_{(2)}(\theta, \gamma) = 0$, $k(\theta, \gamma) = \theta$, $\lambda_1(\theta, \gamma) = -\gamma - \theta^{-1}$, $\lambda_2(\theta, \gamma) = \theta^{-2}$ and $\lambda_3(\theta, \gamma) = -2\theta^{-3}$. Assume that the prior π is a normal density with mean 0 and variance 1. Let θ be known. Since $\pi_{(1)}(\gamma) = -\gamma$, it is seen from (6.6)–(6.8) that the stochastic expansion, the second-order asymptotic mean and asymptotic variance of $T_{B,\theta} = n(\hat{\gamma}_{B,\theta} - \gamma)$ are given by

$$T_{B,\theta} = T_{(1)} - \frac{1}{\theta} - \frac{\gamma}{\theta^2 n} + O_p\left(\frac{1}{n^2}\right),$$

$$E_\gamma(\theta T_{B,\theta}) = -\frac{\gamma}{\theta n} + O\left(\frac{1}{n^2}\right), \quad V_\gamma(\theta T_{B,\theta}) = 1 + O\left(\frac{1}{n^2}\right).$$

Next, let θ be unknown. Since $(\partial/\partial\gamma)logk$ and $\partial^2 k/\partial\theta^2 = 0$, it is seen from (6.11)–(6.13) that

$$T_{B,\hat{\theta}_{ML}} = T_{(1)} - \frac{1}{\theta} + \frac{\hat{U}}{\theta\sqrt{n}} - \frac{\hat{U}^2}{\theta n} - \frac{\gamma}{\theta^2 n} + O_p\left(\frac{1}{n\sqrt{n}}\right),$$

$$E_{\theta,\gamma}[\theta T_{B,\hat{\theta}_{ML}}] = \frac{1}{n}\left(1 - \frac{\gamma}{\theta}\right) + O\left(\frac{1}{n\sqrt{n}}\right),$$

$$V_{\theta,\gamma}(\theta T_{B,\hat{\theta}_{ML}}) = 1 + \frac{1}{n} + O\left(\frac{1}{n\sqrt{n}}\right).$$

Further, it follows from Remark 6.4.2 that the second-order asymptotic loss of $\hat{\gamma}_{B,\hat{\theta}_{ML}}$ relative to $\hat{\gamma}_{B,\theta}$ is given by

$$d_n(\hat{\gamma}_{B,\hat{\theta}_{ML}}, \hat{\gamma}_{B,\theta}) = n\left\{V_{\theta,\gamma}(\theta T_{B,\hat{\theta}_{ML}}) - V_\gamma(\theta T_{B,\theta})\right\} = 1 + o(1)$$

as $n \to \infty$.

Example 6.5.2 **(Lower-truncated normal distribution)** (Continued from Examples 2.7.2 and 4.6.2). Let $c = -\infty$, $d = \infty$, $a(x) = e^{-x^2/2}$, and $u(x) = x$ for $-\infty < \gamma \le x < \infty$ in the density (1.7). Since $b(\theta, \gamma) = \Phi(\theta - \gamma)/\phi(\theta)$ for $\theta \in \Theta = (-\infty, \infty)$, it follows from (4.1), (6.3) and (6.5) that $b_{(1)}(\theta, \gamma) = -\rho(\theta - \gamma)$, $b_{(2)}(\theta, \gamma) = -(\theta - \gamma)\rho(\theta - \gamma) - \rho^2(\theta - \gamma)$, $k(\theta, \gamma) = \rho(\theta - \gamma)$, $\lambda_1(\theta, \gamma) = \theta + \rho(\theta - \gamma)$, $\lambda_2(\theta, \gamma) = 1 - (\theta - \gamma)\rho(\theta - \gamma) - \rho^2(\theta - \gamma)$ and $\lambda_3(\theta, \gamma) = \rho(\theta - \gamma)\{2\rho^2(\theta - \gamma) + 3(\theta - \gamma)\rho(\theta - \gamma) + (\theta - \gamma)^2 - 1\}$, where

$$\Phi(x) := \int_{-\infty}^{x} \phi(t)dt \quad \text{with} \quad \phi(t) := \frac{1}{\sqrt{2\pi}} e^{-t^2/2}$$

and $\rho(t) := \phi(t)/\Phi(t)$. Assume that the prior π is a normal density with mean 0 and variance 1. Let θ be known. Since $\pi_{(1)}(\gamma) = -\gamma$, it is seen from (6.6)–(6.8) that the stochastic expansion, the second-order asymptotic mean, and asymptotic variance of $T_{B,\theta} = n(\hat{\gamma}_{B,\theta} - \gamma)$ are given by

$$T_{B,\theta} = T_{(1)} - \frac{1}{\rho(\theta - \gamma)} + \frac{1}{n}\left\{1 + \frac{\theta - \gamma}{\rho(\theta - \gamma)}\right\} T_{(1)}$$
$$- \frac{1}{\rho^2(\theta - \gamma)n}\{2(\theta - \gamma) + 2\rho(\theta - \gamma) + \gamma\} + O_p\left(\frac{1}{n^2}\right),$$

$$E_{\theta,\gamma}[\rho(\theta - \gamma)T_{B,\theta}] = -\frac{1}{\rho(\theta - \gamma)n}\{2(\theta - \gamma) + 2\rho(\theta - \gamma) + \gamma\} + O\left(\frac{1}{n^2}\right),$$

$$V_{\theta,\gamma}[\rho(\theta - \gamma)T_{B,\theta}] = 1 - \frac{2}{\rho(\theta - \gamma)n}\{\theta - \gamma + \rho(\theta - \gamma)\} + O\left(\frac{1}{n^2}\right).$$

Next, let θ be unknown. Since

$$\frac{\partial^2 k}{\partial \theta^2} = -\rho(\theta-\gamma)\{1-(\theta-\gamma)^2 - 3(\theta-\gamma)\rho(\theta-\gamma) - 2\rho^2(\theta-\gamma)\}$$

and $(\partial/\partial\theta) logk = -(\theta-\gamma) - \rho(\theta-\gamma)$, it is seen from (6.11)–(6.13) that

$$\begin{aligned}
T_{B,\hat{\theta}_{ML}} =& T_{(1)} - \frac{1}{\rho(\theta-\gamma)} - \frac{1}{\sqrt{\lambda_2 n}}\left(1+\frac{\theta-\gamma}{\rho(\theta-\gamma)}\right)\hat{U} + \frac{1}{n}\left(1+\frac{\theta-\gamma}{\rho(\theta-\gamma)}\right)T_{(1)} \\
& - \frac{1}{2\rho(\theta-\gamma)\lambda_2 n}\{1+(\theta-\gamma)^2+(\theta-\gamma)\rho(\theta-\gamma)\}\hat{U}^2 \\
& - \frac{1}{\rho^2(\theta-\gamma)n}\{2(\theta-\gamma+\rho(\theta-\gamma))+\gamma\} + O_p\left(\frac{1}{n\sqrt{n}}\right), \\
E_{\theta,\gamma}[\rho(\theta-\gamma)T_{B,\hat{\theta}_{ML}}] =& -\frac{1}{\rho(\theta-\gamma)n}\{2(\theta-\gamma+\rho(\theta-\gamma))+\gamma\} \\
& + \frac{\{\theta-\gamma+\rho(\theta-\gamma)\}\{\theta-\gamma+2\rho(\theta-\gamma)\}}{2n\{1-(\theta-\gamma)\rho(\theta-\gamma)-\rho^2(\theta-\gamma)\}^2} + O\left(\frac{1}{n\sqrt{n}}\right), \\
V_{\theta,\gamma}(\rho(\theta-\gamma)T_{B,\hat{\theta}_{ML}}) =& 1 - \frac{2\{\theta-\gamma+\rho(\theta-\gamma)\}}{\rho(\theta-\gamma)n} \\
& + \frac{\{\theta-\gamma+\rho(\theta-\gamma)\}^2}{\{1-(\theta-\gamma)\rho(\theta-\gamma)-\rho^2(\theta-\gamma)\}n} + O\left(\frac{1}{n\sqrt{n}}\right),
\end{aligned}$$

where $\lambda_2 = 1-(\theta-\gamma)\rho(\theta-\gamma)-\rho^2(\theta-\gamma)$. Further, it follows from Remark 6.4.2 that the second-order asymptotic loss of $\hat{\gamma}_{B,\hat{\theta}_{ML}}$ relative to $\hat{\gamma}_{B,\theta}$ is given by

$$\begin{aligned}
d_n(\hat{\gamma}_{B,\hat{\theta}_{ML}}, \hat{\gamma}_{B,\theta}) =& n\left\{V_{\theta,\gamma}(\rho(\theta-\gamma)T_{B,\hat{\theta}_{ML}}) - V_\gamma(\rho(\theta-\gamma)T_{B,\theta})\right\} \\
=& \frac{\{\theta-\gamma+\rho(\theta-\gamma)\}^2}{1-(\theta-\gamma)\rho(\theta-\gamma)-\rho^2(\theta-\gamma)} + o(1)
\end{aligned}$$

as $n\to\infty$.

Example 6.5.3 (**Pareto distribution**) (Continued from Examples 2.7.3 and 4.6.3). Let $c=0$, $d=\infty$, $a(x)=1/x$, and $u(x)=-\log x$ for $0<\gamma\le x<\infty$ in the density (1.7). Since $b(\theta,\gamma)=\theta^{-1}\gamma^{-\theta}$ for $\theta\in\Theta=(0,\infty)$, it follows from (4.1), (6.3), and (6.5) that $b_{(1)}(\theta,\gamma)=-\theta/\gamma$, $b_{(2)}(\theta,\gamma)=\theta/\gamma^2$, $k(\theta,\gamma)=\theta/\gamma$, $\lambda_1(\theta,\gamma)=-\theta^{-1}-\log\gamma$, $\lambda_1(\theta,\gamma)=-\theta^{-1}-\log\gamma$, $\lambda_2(\theta,\gamma)=\theta^{-2}$ and $\lambda_3(\theta,\gamma)=-2\theta^{-3}$. Assume that the prior density $\pi(\gamma)$ is $e^{-\gamma}$ for $\gamma>0$. Let θ be known. Since $\pi_{(1)}=-1$, it is seen from (6.6)–(6.8) that the stochastic expansion, the second-order asymptotic mean, and asymptotic variance of $T_{B,\theta}=n(\hat{\gamma}_{B,\theta}-\gamma)$ are given by

$$T_{B,\theta} = T_{(1)} - \frac{\gamma}{\theta} - \frac{1}{\theta n}T_{(1)} + \frac{\gamma^2}{\theta^2 n}\left(\frac{2}{\gamma}-1\right) + O_p\left(\frac{1}{n^2}\right),$$

$$E_\gamma\left(\frac{\gamma}{\theta}T_{B,\theta}\right) = \frac{\gamma}{\theta n}\left(\frac{2}{\gamma} - 1\right) + O\left(\frac{1}{n^2}\right),$$
$$V_\gamma\left(\frac{\gamma}{\theta}T_{B,\theta}\right) = 1 + \frac{2}{\theta n} + O\left(\frac{1}{n^2}\right).$$

Next, let θ be unknown. Since $(\partial/\partial\gamma)logk = -\gamma^{-1}$ and $\partial^2 k/\partial\theta^2 = 0$, it is seen from (6.11)–(6.13) that

$$T_{B,\hat{\theta}_{ML}} = T_{(1)} - \frac{\gamma}{\theta} + \frac{\gamma}{\theta\sqrt{n}}\hat{U} - \frac{1}{\theta n}T_{(1)} - \frac{\gamma}{\theta n}\hat{U}^2 + \frac{\gamma^2}{\theta^2 n}\left(\frac{2}{\gamma} - 1\right) + O_p\left(\frac{1}{n\sqrt{n}}\right),$$
$$E_{\theta,\gamma}\left[\frac{\theta}{\gamma}T_{B,\hat{\theta}_{ML}}\right] = \frac{\gamma}{\theta n}\left(\frac{2}{\gamma} - 1\right) + \frac{1}{n} + O\left(\frac{1}{n\sqrt{n}}\right),$$
$$V_{\theta,\gamma}\left(\frac{\theta}{\gamma}T_{B,\hat{\theta}_{ML}}\right) = 1 + \frac{2}{\theta n} + \frac{1}{n} + O\left(\frac{1}{n\sqrt{n}}\right).$$

Further, it follows from Remark 6.4.2 that the second-order asymptotic loss of $\hat{\gamma}_{B,\hat{\theta}_{ML}}$ relative to $\hat{\gamma}_{B,\theta}$ is given by

$$d_n(\hat{\gamma}_{B,\hat{\theta}_{ML}}, \hat{\gamma}_{B,\theta}) = n\left\{V_{\theta,\gamma}\left(\frac{\theta}{\gamma}T_{B,\hat{\theta}_{ML}}\right) - V_\gamma\left(\frac{\theta}{\gamma}T_{B,\theta}\right)\right\} = 1 + o(1)$$

as $n \to \infty$.

Example 6.5.4 **(Lower-truncated beta distribution)** (Continued from Examples 2.7.4 and 4.6.4). Let $c = 0, d = 1, a(x) = x^{-1}$, and $u(x) = \log x$ for $0 < \gamma \le x < 1$ in the density (1.7). Since $b(\theta,\gamma) = \theta^{-1}(1-\gamma^\theta)$ for $\theta \in \Theta = (0,\infty)$, it follows from (4.1), (6.3), and (6.5) that

$$k(\theta,\gamma) = \theta\gamma^{\theta-1}/(1-\gamma^\theta), \quad \frac{\partial}{\partial\gamma}\log k(\theta,\gamma) = \frac{\theta-1}{\gamma} + \frac{\theta\gamma^{\theta-1}}{1-\gamma^\theta}, \tag{6.14}$$
$$\lambda_1(\theta,\gamma) = -\frac{1}{\theta} - \frac{(\log\gamma)\gamma^\theta}{1-\gamma^\theta}, \quad \lambda_2(\theta,\gamma) = \frac{1}{\theta^2} - \frac{(\log\gamma)^2\gamma^\theta}{(1-\gamma^\theta)^2}, \tag{6.15}$$
$$\lambda_3(\theta,\gamma) = -\frac{2}{\theta^3} - (\log\gamma)^3\gamma^\theta\frac{1+\gamma^\theta}{(1-\gamma^\theta)^3}. \tag{6.16}$$

Assume that the prior density is

$$\pi(\gamma) = \begin{cases}\alpha\gamma^{\alpha-1} & \text{for } 0 < \gamma < 1,\\ 0 & \text{otherwise,}\end{cases}$$

where α is positive and known. Let θ be known and $\hat{\gamma}_{B,\theta}$ be the Bayes estimator of γ. Since $\pi_{(1)}(\gamma) = (\alpha-1)/\gamma$ for $\gamma > 0$, it is seen from (6.6)–(6.8) that the stochastic expansion, second-order asymptotic mean, and asymptotic variance of $T_{B,\theta} = n(\hat{\gamma}_{B,\theta} - \gamma)$ are given by

$$T_{B,\theta} = T_{(1)} - \frac{1-\gamma^\theta}{\theta\gamma^{\theta-1}} + \frac{\theta - (1-\gamma^\theta)}{\theta\gamma^\theta n} T_{(1)} + \frac{1-\gamma^\theta}{\theta^2\gamma^{2\theta-1}n}\{(\alpha+1)(1-\gamma^\theta) - 2\theta\} + O_p\left(\frac{1}{n^2}\right),$$

$$E_\gamma(kT_{B,\theta}) = \frac{1}{\theta\gamma^\theta n}\{\alpha - 2\theta + 1 - (\alpha+1)\gamma^\theta\} + O\left(\frac{1}{n^2}\right),$$

$$V_\gamma(kT_{B,\theta}) = 1 - \frac{2}{\theta\gamma^\theta n}\{\theta - (1-\gamma^\theta)\} + O\left(\frac{1}{n^2}\right),$$

where $k = k(\theta, \gamma)$. Next, let θ be unknown and $\hat{\gamma}_{B,\hat{\theta}_{ML}}$ be the Bayes estimator plugging $\hat{\theta}_{ML}$ in θ of $\hat{\gamma}_{B,\theta}$. Since

$$\frac{\partial\lambda_1}{\partial\gamma} = -\frac{\gamma^{\theta-1}}{(1-\gamma^\theta)^2}(1 - \gamma^\theta - \theta\log\gamma),$$

$$\frac{\partial k}{\partial\theta} = \frac{\gamma^{\theta-1}}{1-\gamma^\theta}\left(1 + \frac{\theta\log\gamma}{1-\gamma^\theta}\right), \quad \frac{\partial^2 k}{\partial\theta^2} = \frac{(\log\gamma)\theta\gamma^{\theta-1}}{(1-\gamma^\theta)^2}\left\{\frac{2}{\theta} + \frac{(\log\gamma)(1+\gamma^\theta)}{1-\gamma^\theta}\right\},$$

in a similar way to the above, from (6.11)–(6.13) and (6.14)–(6.16), we obtain the stochastic expansion, second-order asymptotic mean, and asymptotic variance of $T_{B,\hat{\theta}_{ML}}$. In particular

$$V_{\theta,\gamma}(kT_{B,\hat{\theta}_{ML}}) = 1 - \frac{2}{\theta\gamma^\theta n}\{\theta - (1-\gamma^\theta)\} + \frac{1}{\lambda_2 n}\left(\frac{1}{\theta} + \frac{\log\gamma}{1-\gamma^\theta}\right)^2 + O\left(\frac{1}{n\sqrt{n}}\right),$$

where λ_2 is given by (6.15). Further, it follows from Remark 6.4.2 that the second-order asymptotic loss of $\hat{\gamma}_{B,\hat{\theta}_{ML}}$ relative to $\hat{\gamma}_{B,\theta}$ is given by

$$d_n(\hat{\gamma}_{B,\hat{\theta}_{ML}}, \hat{\gamma}_{B,\theta}) = n\{V_{\theta,\gamma}(kT_{B,\hat{\theta}_{ML}}) - V_\gamma(kT_{B,\theta})\} = \frac{1}{\lambda_2}\left(\frac{1}{\theta} + \frac{\log\gamma}{1-\gamma^\theta}\right)^2 + o(1)$$

as $n \to \infty$.

Example 6.5.5 (**Lower-truncated Erlang distribution**) (Continued from Examples 2.7.5 and 4.6.5). Let $c = 0$, $d = \infty$, $a(x) = x^{j-1}$, and $u(x) = -x$ for $0 < \gamma \le x < \infty$ in the density (1.7), where $j = 1, 2, \ldots$. Note that the density is truncated exponential for $j = 1$. For each $j = 1, 2, \ldots$, $b_j(\theta, \gamma) := \int_\gamma^\infty x^{j-1}e^{-\theta x}dx$ for $\theta \in \Theta = (0, \infty)$. Since $\partial b_j/\partial\theta = -b_{j+1}$ $(j = 1, 2, \ldots)$, it follows from (2.1), (6.3), and (6.5) that, for each $j = 1, 2, \ldots$

$$k_j(\theta,\gamma) := \frac{a(\gamma)e^{\theta u(\gamma)}}{b_j(\theta,\gamma)} = \frac{\gamma^{j-1}e^{-\theta\gamma}}{b_j(\theta,\gamma)}, \quad \frac{\partial}{\partial\gamma}\log k_j(\theta,\gamma) = \frac{j-1}{\gamma} - \theta + k_j(\theta,\gamma), \tag{6.17}$$

$$\lambda_{j1} := \frac{\partial}{\partial\theta}\log b_j(\theta,\gamma) = -\frac{b_{j+1}}{b_j}, \quad \lambda_{j2} := \frac{\partial^2}{\partial\theta^2}\log b_j(\theta,\gamma) = \frac{b_{j+2}}{b_j} - \left(\frac{b_{j+1}}{b_j}\right)^2, \tag{6.18}$$

$$\lambda_{j3} := \frac{\partial^3}{\partial\theta^3}\log b_j(\theta,\gamma) = -\frac{b_{j+3}}{b_j} + \frac{3b_{j+1}b_{j+2}}{b_j^2} - 2\left(\frac{b_{j+1}}{b_j}\right)^3, \tag{6.19}$$

where $b_j = b_j(\theta,\gamma)$ $(j = 1, 2, \ldots)$. Assume that the prior density is

$$\pi(\gamma) = \begin{cases} \frac{1}{\Gamma(\alpha)}\gamma^{\alpha-1}e^{-\gamma} & \text{for } \gamma > 0, \\ 0 & \text{otherwise,} \end{cases}$$

where α is positive and known. Let θ be known and, for each $j = 1, 2, \ldots$, let $\hat{\gamma}_{B,\theta}^{(j)}$ be the Bayes estimator of γ. Since $\pi_{(1)}(\gamma) = \{(\alpha-1)/\gamma\} - 1$ for $\gamma > 0$, it is seen from (6.6)–(6.8) that, for each $j = 1, 2, \ldots$, the stochastic expansion, second-order asymptotic mean and asymptotic variance of $T_{B,\theta}^{(j)} = n(\hat{\gamma}_{B,\theta}^{(j)} - \gamma)$ are given by

$$\begin{aligned} T_{B,\theta}^{(j)} = T_{(1)} &- \frac{b_j}{\gamma^{j-1}e^{-\theta\gamma}} + \frac{b_j T_{(1)}}{\gamma^{j-1}e^{-\theta\gamma}n}\left(\frac{j-1}{\gamma} - \theta + \frac{\gamma^{j-1}e^{-\theta\gamma}}{b_j}\right) \\ &- \frac{b_j^2}{\gamma^{2(j-1)}e^{-2\theta\gamma}n}\left\{2\left(\frac{j-1}{\gamma} - \theta + \frac{\gamma^{j-1}e^{-\theta\gamma}}{b_j}\right) - \frac{\alpha-1}{\gamma} + 1\right\} + O_p\left(\frac{1}{n^2}\right), \\ E_\gamma(k_j T_{B,\theta}^{(j)}) &= -\frac{b_j}{\gamma^{j-1}e^{-\theta\gamma}n}\left\{2\left(\frac{j-1}{\gamma} - \theta + \frac{\gamma^{j-1}e^{-\theta\gamma}}{b_j}\right) - \frac{\alpha-1}{\gamma} + 1\right\} + O\left(\frac{1}{n^2}\right), \\ V_\gamma(k_j T_{B,\theta}^{(j)}) &= 1 - \frac{2b_j}{\gamma^{j-1}e^{-\theta\gamma}n}\left(\frac{j-1}{\gamma} - \theta + \frac{\gamma^{j-1}e^{-\theta\gamma}}{b_j}\right) + O\left(\frac{1}{n^2}\right), \end{aligned}$$

where $k_j = k_j(\theta,\gamma)$ and $b_j = b_j(\theta,\gamma)$. Next, let θ be unknown and, for each $j = 1, 2, \ldots$, let $\hat{\gamma}_{B,\hat{\theta}_{ML}}^{(j)}$ be the Bayes estimator plugging $\hat{\theta}_{ML}$ in θ of $\hat{\gamma}_{B,\theta}^{(j)}$. Since, for each $j = 1, 2, \ldots$,

$$\begin{aligned} \frac{\partial\lambda_{1j}}{\partial\gamma} &= \frac{\gamma^{j-1}e^{-\theta\gamma}}{b_j}\left(\frac{b_{j+1}}{b_j} + \gamma\right), \quad \frac{\partial k_j}{\partial\theta} = \frac{\gamma^{j-1}e^{-\theta\gamma}}{b_j}\left(\frac{b_{j+1}}{b_j} - \gamma\right), \\ \frac{\partial^2 k_j}{\partial\theta^2} &= \frac{\gamma^{j-1}e^{-\theta\gamma}}{b_j}\left\{\left(\frac{b_{j+1}}{b_j}\right)^2 + \left(\frac{b_{j+1}}{b_j} - \gamma\right)^2 - \frac{b_{j+2}}{b_j}\right\}, \end{aligned}$$

in a similar way to the above, from (6.11)–(6.13) and (6.17)–(6.19), we obtain the stochastic expansion, second-order asymptotic mean, and asymptotic variance of $T_{B,\hat{\theta}_{ML}}^{(j)}$ for each $j = 1, 2, \ldots$. In particular,

$$V_{\theta,\gamma}(k_j T^{(j)}_{B,\hat{\theta}_{ML}}) = 1 - \frac{2b_j}{\gamma^{j-1}e^{-\theta\gamma}n}\left(\frac{j-1}{\gamma} - \theta + \frac{\gamma^{j-1}e^{-\theta\gamma}}{b_j}\right) + \frac{b_j^2}{(b_j b_{j+2} - b_{j+1}^2)n}\left(\frac{b_{j+1}}{b_j} - \gamma\right)^2 + O\left(\frac{1}{n\sqrt{n}}\right).$$

Further, it follows from Remark 6.4.2 that, for each $j = 1, 2, \ldots,$ the second-order asymptotic loss of $\hat{\gamma}^{(j)}_{B,\hat{\theta}_{ML}}$ relative to $\hat{\gamma}^{(j)}_{B,\theta}$ is given by

$$\begin{aligned} d_n(\hat{\gamma}^{(j)}_{B,\hat{\theta}_{ML}}, \hat{\gamma}^{(j)}_{B,\theta}) &= n\left\{V_{\theta,\gamma}\left(k_j T^{(j)}_{B,\hat{\theta}_{ML}}\right) - V_{\gamma}\left(k_j T^{(j)}_{B,\theta}\right)\right\} \\ &= \frac{b_j^2}{(b_j b_{j+2} - b_{j+1}^2)n}\left(\frac{b_{j+1}}{b_j} - \gamma\right)^2 + o(1) \end{aligned}$$

as $n \to \infty$.

A similar example to the above in the lower-truncated lognormal case in Example 2.7.6 is reduced to the lower-truncated normal one in Example 6.5.2. In the case, the standard normal distribution may be taken as a prior one.

6.6 Concluding Remarks

For a oTEF $\mathscr{P}_o$ of distributions with a truncation parameter γ and a natural parameter θ, we considered the estimation problem on γ in the presence of θ as a nuisance parameter from the Bayesian viewpoint. Under the smooth prior density $\pi(\gamma)$ of γ and a quadratic loss, the stochastic expansions of the Bayes estimator $\hat{\gamma}_{B,\theta}$ (i.e., (6.1)) of γ when θ is known and the Bayes estimator $\hat{\gamma}_{B,\hat{\theta}_{ML}}$ (i.e., (6.10)) plugging the MLE $\hat{\theta}_{ML}$ in θ of $\hat{\gamma}_{B,\theta}$ when θ is unknown were derived, which led to the fact that the asymptotic means of $T_{B,\theta} = n(\hat{\gamma}_{B,\theta} - \gamma)$ and $T_{B,\hat{\theta}_{ML}} = n(\hat{\gamma}_{B,\hat{\theta}_{ML}} - \gamma)$ depended on the prior π, but their second-order asymptotic variances were independent of it. In the previous discussion, we adopted a partial Bayesian approach; to be precise, we chose the combined Bayesian-frequentist approach. Indeed, since a density (1.7) in the oTEF $\mathscr{P}_o$ has a truncation point γ, it is considered to be helpful to obtain some information through a prior on γ. On the other hand, a natural parameter θ in $\mathscr{P}_o$ is in the same situation as in a regular exponential family, hence the maximum likelihood method based on the likelihood equation is useful for estimating θ. Hence, it seems to be natural to plug the MLE in θ of the Bayes estimator of γ for known θ. But, taking a pure Bayesian approach, one may obtain the Bayes estimator with respect to a prior on (θ, γ). It seems to be interesting to compare it with our estimator. For a tTEF $\mathscr{P}_t$ of distributions, we may also obtain the Bayes estimator with respect to a prior on (θ, γ, ν) and compare it with the partial Bayes estimator.

6.7 Appendix E

The proof of Theorem 6.3.1 Since, by the Taylor expansion,

$$b\left(\theta, \gamma + \frac{u}{n}\right)$$
$$= b(\theta, \gamma)\left\{1 + \left(\frac{\partial}{\partial \gamma} \log b(\theta, \gamma)\right)\frac{u}{n} + \frac{1}{2b(\theta, \gamma)}\left(\frac{\partial^2 b(\theta, \gamma)}{\partial \gamma^2}\right)\frac{u^2}{n^2} + O\left(\frac{1}{n^3}\right)\right\}$$

as $n \to \infty$, it follows that

$$b^n\left(\theta, \gamma + \frac{u}{n}\right) = b^n(\theta, \gamma) \exp\left\{b_{(1)}u - \frac{1}{2n}b_{(1)}^2 u^2 + \frac{1}{2bn}\left(\frac{\partial^2 b}{\partial \gamma^2}\right)u^2 + O\left(\frac{1}{n^2}\right)\right\}, \tag{6.20}$$

where $b = b(\theta, \gamma)$ and $b_{(j)} = b_{(j)}(\theta, \gamma)$. Here

$$\frac{1}{b(\theta, \gamma)}\frac{\partial^2}{\partial \gamma^2}b(\theta, \gamma) = b_{(2)} + b_{(1)}^2. \tag{6.21}$$

Substituting (6.21) into (6.20), we have

$$b^n\left(\theta, \gamma + \frac{u}{n}\right) = b^n(\theta, \gamma)e^{b_{(1)}u}\left\{1 + \frac{b_{(2)}}{2n}u^2 + O\left(\frac{1}{n^2}\right)\right\}. \tag{6.22}$$

From (6.4), we have

$$\pi\left(\gamma + \frac{u}{n}\right) = \pi(\gamma)\left\{1 + \pi_{(1)}\frac{u}{n} + O\left(\frac{1}{n^2}\right)\right\}, \tag{6.23}$$

where $\pi_{(1)} = \pi_{(1)}(\gamma)$. From (6.22) and (6.23), we obtain

$$\frac{\pi(\gamma + (u/n))}{b^n(\theta, \gamma + (u/n))} = \frac{\pi(\gamma)e^{-b_{(1)}u}}{b^n(\theta, \gamma)}\left\{1 - \frac{b_{(2)}}{2n}u^2 + \frac{\pi_{(1)}}{n}u + O\left(\frac{1}{n^2}\right)\right\}. \tag{6.24}$$

Putting

$$I_j := \int_{\tau_n}^{T_{(1)}} u^j e^{ku} du \quad (j = 0, 1, 2, 3),$$

we have from (6.5) and (6.24)

$$\int_{\tau_n}^{T_{(1)}} \frac{\pi(\gamma+(u/n))}{b^n(\theta,\gamma+(u/n))}du = \frac{\pi(\gamma)}{b^n(\theta,\gamma)}\left\{I_0 - \frac{b_{(2)}}{2n}I_2 + \frac{\pi_{(1)}}{n}I_1 + O_p\left(\frac{1}{n^2}\right)\right\}, \tag{6.25}$$

$$\int_{\tau_n}^{T_{(1)}} \frac{u\pi(\gamma+(u/n))}{b^n(\theta,\gamma+(u/n))}du = \frac{\pi(\gamma)}{b^n(\theta,\gamma)}\left\{I_1 - \frac{b_{(2)}}{2n}I_3 + \frac{\pi_{(1)}}{n}I_2 + O_p\left(\frac{1}{n^2}\right)\right\}. \tag{6.26}$$

Here, the remainder terms in (6.25) and (6.26) are guaranteed to be order $O_p(n^{-2})$, since the distribution with a density (1.7) belongs to a oTEF $\mathscr{P}_o$ with a normalizing factor $b(\theta,\gamma)$ based on $a(\cdot)$ and $u(\cdot)$ which are functions of class C^3. From (6.25) and (6.26), we obtain

$$\int_{\tau_n}^{T_{(1)}} \frac{u\pi(\gamma+(u/n))}{b^n(\theta,\gamma+(u/n))}du \Bigg/ \int_{\tau_n}^{T_{(1)}} \frac{\pi(\gamma+(u/n))}{b^n(\theta,\gamma+(u/n))}du$$
$$= \frac{I_1}{I_0} + \frac{b_{(2)}}{2n}\left(\frac{I_1I_2}{I_0^2} - \frac{I_3}{I_0}\right) - \frac{\pi_{(1)}}{n}\left\{\left(\frac{I_1}{I_0}\right)^2 - \frac{I_2}{I_0}\right\} + O_p\left(\frac{1}{n^2}\right). \tag{6.27}$$

Substituting (6.27) into (6.2), we have

$$T_{B,\theta} := n(\hat{\gamma}_{B,\theta} - \gamma) = \frac{I_1}{I_0} + \frac{b_{(2)}}{2n}\left(\frac{I_1I_2}{I_0^2} - \frac{I_3}{I_0}\right) - \frac{\pi_{(1)}}{n}\left\{\left(\frac{I_1}{I_0}\right)^2 - \frac{I_2}{I_0}\right\} + O_p\left(\frac{1}{n^2}\right). \tag{6.28}$$

Since

$$I_0 = \int_{\tau_n}^{T_{(1)}} e^{ku}du = \frac{1}{k}e^{kT_{(1)}} + O_p\left(e^{k\tau_n}\right),$$
$$I_1 = \int_{\tau_n}^{T_{(1)}} ue^{ku}du = \frac{1}{k}e^{kT_{(1)}}\left(T_{(1)} - \frac{1}{k}\right) + O_p\left(ne^{k\tau_n}\right),$$
$$I_2 = \int_{\tau_n}^{T_{(1)}} u^2e^{ku}du = \frac{1}{k}e^{kT_{(1)}}\left\{\left(T_{(1)} - \frac{1}{k}\right)^2 + \frac{1}{k^2}\right\} + O_p\left(n^2e^{k\tau_n}\right),$$
$$I_3 = \int_{\tau_n}^{T_{(1)}} u^3e^{ku}du = \frac{1}{k}e^{kT_{(1)}}\left\{\left(T_{(1)} - \frac{1}{k}\right)^3 + \frac{3}{k^2}\left(T_{(1)} - \frac{1}{k}\right) - \frac{2}{k^3}\right\} + O_p\left(n^3e^{k\tau_n}\right)$$

as $n \to \infty$, it follows that

$$\frac{I_1}{I_0} = T_{(1)} - \frac{1}{k} + O_p\left(e^{k(\tau_n - T_{(1)})}\right), \tag{6.29}$$

$$\frac{I_2}{I_0} = \left(T_{(1)} - \frac{1}{k}\right)^2 + \frac{1}{k^2} + O_p\left(n^2 e^{k(\tau_n - T_{(1)})}\right), \tag{6.30}$$

$$\frac{I_3}{I_0} = \left(T_{(1)} - \frac{1}{k}\right)^3 + \frac{3}{k^2}\left(T_{(1)} - \frac{1}{k}\right) - \frac{2}{k^3} + O_p\left(n^3 e^{k(\tau_n - T_{(1)})}\right), \tag{6.31}$$

hence, by (6.28)

$$T_{B,\theta} = T_{(1)} - \frac{1}{k} - \frac{b_{(2)}}{k^2 n} T_{(1)} + \frac{1}{k^2 n}\left(\frac{2b_{(2)}}{k} + \pi_{(1)}\right) + O_p\left(\frac{1}{n^2}\right). \tag{6.32}$$

Since by (6.3) and (6.5)

$$\frac{b_{(2)}}{k} = \frac{1}{k}\left(\frac{\partial^2}{\partial \gamma^2} \log b\right) = -\frac{1}{k}\left(\frac{\partial k}{\partial \gamma}\right) = -\frac{\partial}{\partial \gamma} \log k, \tag{6.33}$$

it follows from (6.32) and (6.33) that

$$T_{B,\theta} = T_{(1)} - \frac{1}{k} + \frac{1}{kn}\left(\frac{\partial}{\partial \gamma} \log k\right) T_{(1)} - \frac{1}{k^2 n}\left\{2\left(\frac{\partial}{\partial \gamma} \log k\right) - \pi_{(1)}\right\} + O_p\left(\frac{1}{n^2}\right),$$

which implies that (6.6) holds. From (6.6) and Theorem 4.3.1, we have

$$T_{B,\theta} = T_{(1)}^* - \frac{1}{k^2 n}\left\{2\left(\frac{\partial}{\partial \gamma} \log k\right) - \pi_{(1)}\right\} + O_p\left(\frac{1}{n^2}\right),$$

hence, by (4.7) and (4.8), we obtain (6.7) and (6.8). Thus, we complete the proof.

The proof of Theorem 6.4.1 The Bayes estimator (6.10) plugging $\hat{\theta}_{ML}$ in θ of $\hat{\gamma}_{B,\theta}$ when θ is unknown is expressed by

$$\hat{\gamma}_{B,\hat{\theta}_{ML}}(\boldsymbol{X}) = \gamma + \frac{1}{n}\left(\int_{\tau_n}^{T_{(1)}} \frac{u\pi(\gamma + (u/n))}{b^n\left(\hat{\theta}_{ML}, \gamma + (u/n)\right)} du \Big/ \int_{\tau_n}^{T_{(1)}} \frac{\pi(\gamma + (u/n))}{b^n\left(\hat{\theta}_{ML}, \gamma + (u/n)\right)} du\right), \tag{6.34}$$

where $\tau_n = n(c - r)$. Since $\hat{\theta}_{ML} = \theta + (\hat{U}/\sqrt{\lambda_2 n})$ and

$$\frac{1}{b(\theta,\gamma)} \frac{\partial^2}{\partial \theta^2} b(\theta,\gamma) = \lambda_2 + \lambda_1^2, \quad \frac{1}{b(\theta,\gamma)} \frac{\partial^3}{\partial \theta^3} b(\theta,\gamma) = \lambda_3 + 3\lambda_1\lambda_2 + \lambda_1^3,$$

we have by the Taylor expansion

$$
\begin{aligned}
& b\left(\theta+\frac{\hat{U}}{\sqrt{\lambda_2 n}}, \gamma+\frac{u}{n}\right) \\
&= b(\theta, \gamma)\left\{1+\frac{\lambda_1}{\sqrt{\lambda_2 n}}\hat{U}+\frac{b_{(1)}}{n}u+\frac{\lambda_2+\lambda_1^2}{2\lambda_2 n}\hat{U}^2+\frac{1}{b}\left(\frac{\partial^2 b}{\partial\theta\partial\gamma}\right)\frac{\hat{U}u}{\sqrt{\lambda_2 n}\sqrt{n}}\right. \\
&\quad +\frac{\lambda_3+3\lambda_1\lambda_2+\lambda_1^3}{6\lambda_2^{3/2}n\sqrt{n}}\hat{U}^3+\frac{1}{2b\lambda_2 n^2}\left(\frac{\partial^3 b}{\partial\theta^2\partial\gamma}\right)\hat{U}^2 u \\
&\quad \left.+\frac{1}{2bn^2}\left(\frac{\partial^2 b}{\partial\gamma^2}\right)u^2+\frac{1}{24b\lambda_2^2 n^2}\left(\frac{\partial^4 b}{\partial\theta^4}\right)\hat{U}^4+O\left(\frac{1}{n^2\sqrt{n}}\right)\right\},
\end{aligned}
$$

hence

$$
\begin{aligned}
& b^n\left(\theta+\frac{\hat{U}}{\sqrt{\lambda_2 n}}, \gamma+\frac{u}{n}\right) \\
&= b^n(\theta, \gamma)\left[\exp\left\{\frac{\lambda_1\sqrt{n}}{\sqrt{\lambda_2}}\hat{U}+\frac{1}{2}\hat{U}^2+\frac{\lambda_3}{6\lambda_2^{3/2}\sqrt{n}}\hat{U}^3\right.\right. \\
&\quad \left.\left.+\frac{1}{24b\lambda_2^2 n}\left(\frac{\partial^4 b}{\partial\theta^4}\right)\hat{U}^4-\frac{\lambda_1^4+6\lambda_1^2\lambda_2+4\lambda_1\lambda_3+3\lambda_2^2}{24\lambda_2^2 n}\hat{U}^4\right\}\right] \\
&\quad \cdot e^{b_{(1)}u}\exp\left\{\frac{\hat{U}}{b\sqrt{\lambda_2 n}}\left(\frac{\partial^2 b}{\partial\theta\partial\gamma}\right)u+\frac{b_{(2)}}{2n}u^2+\frac{\hat{U}^2}{2b\lambda_2 n}\left(\frac{\partial^3 b}{\partial\theta^2\partial\gamma}\right)u-\frac{\lambda_1 b_{(1)}\hat{U}}{\sqrt{\lambda_2 n}}u\right. \\
&\quad \left.-\frac{b_{(1)}(\lambda_2+\lambda_1^2)\hat{U}^2}{2\lambda_2 n}u-\frac{\lambda_1}{b\lambda_2 n}\left(\frac{\partial^2 b}{\partial\theta\partial\gamma}\right)\hat{U}^2 u+\frac{b_{(1)}\lambda_1^2}{\lambda_2 n}\hat{U}^2 u+O_p\left(\frac{1}{n\sqrt{n}}\right)\right\} \\
&=: b^n(\theta, \gamma)[\exp\{Q\}]e^{b_{(1)}u}\exp\left\{\frac{\hat{U}}{b\sqrt{\lambda_2 n}}\left(\frac{\partial^2 b}{\partial\theta\partial\gamma}\right)u+\frac{b_{(2)}}{2n}u^2+\frac{\hat{U}^2}{2b\lambda_2 n}\left(\frac{\partial^3 b}{\partial\theta^2\partial\gamma}\right)u\right. \\
&\quad \left.-\frac{\lambda_1 b_{(1)}\hat{U}}{\sqrt{\lambda_2 n}}u-\frac{b_{(1)}(\lambda_2-\lambda_1^2)\hat{U}^2}{2\lambda_2 n}u-\frac{\lambda_1}{b\lambda_2 n}\left(\frac{\partial^2 b}{\partial\theta\partial\gamma}\right)\hat{U}^2 u+O_p\left(\frac{1}{n\sqrt{n}}\right)\right\},
\end{aligned}
\tag{6.35}
$$

where Q is independent of u. From (6.23) and (6.35), we have

$$
\begin{aligned}
& \pi\left(\gamma+\frac{u}{n}\right)\Big/ b^n\left(\theta+\frac{\hat{U}}{\sqrt{\lambda_2 n}}, \gamma+\frac{u}{n}\right) \\
&= b^{-n}(\theta, \gamma)[\exp\{-Q\}]\pi(\gamma)e^{-b_{(1)}u}\left[1-\frac{\hat{U}}{b\sqrt{\lambda_2 n}}\left(\frac{\partial^2 b}{\partial\theta\partial\gamma}\right)u+\frac{\lambda_1 b_{(1)}}{\sqrt{\lambda_2 n}}\hat{U}u-\frac{b_{(2)}}{2n}u^2\right.
\end{aligned}
$$

$$+\frac{b_{(1)}(\lambda_2-\lambda_1^2)}{2\lambda_2 n}\hat{U}^2u-\frac{\hat{U}^2}{2b\lambda_2 n}\left(\frac{\partial^3 b}{\partial\theta^2\partial\gamma}\right)u+\frac{\lambda_1}{b\lambda_2 n}\left(\frac{\partial^2 b}{\partial\theta\partial\gamma}\right)\hat{U}^2u+\frac{\pi_{(1)}}{n}u$$
$$+\frac{1}{2}\left\{\frac{1}{b^2}\left(\frac{\partial^2 b}{\partial\theta\partial\gamma}\right)^2\frac{1}{\lambda_2 n}\hat{U}^2u^2-\frac{2\lambda_1 b_{(1)}\hat{U}^2}{b\lambda_2 n}\left(\frac{\partial^2 b}{\partial\theta\partial\gamma}\right)u^2+\frac{\lambda_1^2 b_{(1)}^2}{\lambda_2 n}\hat{U}^2u^2\right\}$$
$$+O_p\left(\frac{1}{n\sqrt{n}}\right)\Bigg],$$

hence

$$\int_{\tau_n}^{T_{(1)}}\frac{\pi\left(\gamma+(u/n)\right)}{b^n\left(\theta+(\hat{U}/\sqrt{\lambda_2 n}),\gamma+(u/n)\right)}du$$
$$=b^{-n}(\theta,\gamma)[\exp\{-Q\}]\pi(\gamma)\Bigg[I_0+\frac{\hat{U}}{\sqrt{\lambda_2 n}}\left\{\lambda_1 b_{(1)}-\frac{1}{b}\left(\frac{\partial^2 b}{\partial\theta\partial\gamma}\right)\right\}I_1$$
$$-\frac{b_{(2)}}{2n}I_2+\frac{b_{(1)}(\lambda_2-\lambda_1^2)\hat{U}^2}{2\lambda_2 n}I_1-\frac{1}{\lambda_2 n}\left\{\frac{1}{2b}\left(\frac{\partial^3 b}{\partial\theta^2\partial\gamma}\right)-\frac{\lambda_1}{b}\left(\frac{\partial^2 b}{\partial\theta\partial\gamma}\right)\right\}\hat{U}^2I_1$$
$$+\frac{\hat{U}^2I_2}{\lambda_2 n}\left\{\frac{1}{2b^2}\left(\frac{\partial^2 b}{\partial\theta\partial\gamma}\right)^2-\frac{\lambda_1 b_{(1)}}{b}\left(\frac{\partial^2 b}{\partial\theta\partial\gamma}\right)+\frac{1}{2}\lambda_1^2 b_{(1)}^2\right\}+\frac{\pi_{(1)}}{n}I_1+O_p\left(\frac{1}{n\sqrt{n}}\right)\Bigg], \tag{6.36}$$

$$\int_{\tau_n}^{T_{(1)}}\frac{u\pi\left(\gamma+(u/n)\right)}{b^n\left(\theta+(\hat{U}/\sqrt{\lambda_2 n}),\gamma+(u/n)\right)}du$$
$$=b^{-n}(\theta,\gamma)[\exp\{-Q\}]\pi(\gamma)\Bigg[I_1+\frac{\hat{U}}{\sqrt{\lambda_2 n}}\left\{\lambda_1 b_{(1)}-\frac{1}{b}\left(\frac{\partial^2 b}{\partial\theta\partial\gamma}\right)\right\}I_2$$
$$-\frac{b_{(2)}}{2n}I_3+\frac{b_{(1)}(\lambda_2-\lambda_1^2)\hat{U}^2}{2\lambda_2 n}I_2-\frac{1}{\lambda_2 n}\left\{\frac{1}{2b}\left(\frac{\partial^3 b}{\partial\theta^2\partial\gamma}\right)-\frac{\lambda_1}{b}\left(\frac{\partial^2 b}{\partial\theta\partial\gamma}\right)\right\}\hat{U}^2I_2$$
$$+\frac{\hat{U}^2I_3}{\lambda_2 n}\left\{\frac{1}{2b^2}\left(\frac{\partial^2 b}{\partial\theta\partial\gamma}\right)^2-\frac{\lambda_1 b_{(1)}}{b}\left(\frac{\partial^2 b}{\partial\theta\partial\gamma}\right)+\frac{1}{2}\lambda_1^2 b_{(1)}^2\right\}+\frac{\pi_{(1)}}{n}I_2+O_p\left(\frac{1}{n\sqrt{n}}\right)\Bigg]. \tag{6.37}$$

Here, the remainder terms in (6.36) and (6.37) are guaranteed to be order $O_p(n^{-3/2})$, since the distribution with a density (1.7) belongs to a oTEF $\mathscr{P}_o$ with a normalizing factor $b(\theta,\gamma)$ based on $a(\cdot)$ and $u(\cdot)$ which are functions of class C^3. Substituting (6.36) and (6.37) into (6.34), we have

$$n(\hat{\gamma}_{B,\hat{\theta}_{ML}}-\gamma)$$
$$=\frac{I_1}{I_0}+\frac{\hat{U}}{\sqrt{\lambda_2 n}}\left\{\lambda_1 b_{(1)}-\frac{1}{b}\left(\frac{\partial^2 b}{\partial\theta\partial\gamma}\right)\right\}\left\{\frac{I_2}{I_0}-\left(\frac{I_1}{I_0}\right)^2\right\}-\frac{1}{2n}b_{(2)}\left(\frac{I_3}{I_0}-\frac{I_1I_2}{I_0^2}\right)$$
$$+\frac{b_{(1)}(\lambda_2-\lambda_1^2)}{2\lambda_2 n}\hat{U}^2\left\{\frac{I_2}{I_0}-\left(\frac{I_1}{I_0}\right)^2\right\}-\frac{\hat{U}^2}{\lambda_2 n}\left\{\frac{1}{2b}\left(\frac{\partial^3 b}{\partial\theta^2\partial\gamma}\right)-\frac{\lambda_1}{b}\left(\frac{\partial^2 b}{\partial\theta\partial\gamma}\right)\right\}$$

$$\cdot\left\{\frac{I_2}{I_0}-\left(\frac{I_1}{I_0}\right)^2\right\}+\frac{\hat{U}^2}{\lambda_2 n}\left\{\frac{1}{2b^2}\left(\frac{\partial^2 b}{\partial\theta\partial\gamma}\right)^2-\frac{\lambda_1 b_{(1)}}{b}\left(\frac{\partial^2 b}{\partial\theta\partial\gamma}\right)+\frac{1}{2}\lambda_1^2 b_{(1)}^2\right\}$$

$$\cdot\left(\frac{I_3}{I_0}-\frac{I_1 I_2}{I_0^2}\right)+\frac{\pi_{(1)}}{n}\left\{\frac{I_2}{I_0}-\left(\frac{I_1}{I_0}\right)^2\right\}+\frac{\hat{U}^2}{\lambda_2 n}\left\{\lambda_1 b_{(1)}-\frac{1}{b}\left(\frac{\partial^2 b}{\partial\theta\partial\gamma}\right)\right\}^2\left(\frac{I_1}{I_0}\right)^3$$

$$-\frac{\hat{U}^2}{\lambda_2 n}\left\{\lambda_1 b_{(1)}-\frac{1}{b}\left(\frac{\partial^2 b}{\partial\theta\partial\gamma}\right)\right\}^2\frac{I_1 I_2}{I_0^2}+O_p\left(\frac{1}{n\sqrt{n}}\right). \tag{6.38}$$

From (6.29)–(6.31) and (6.38), we obtain

$$\begin{aligned}
&n(\hat{\gamma}_{B,\hat{\theta}_{ML}}-\gamma)\\
&=T_{(1)}+\frac{1}{b_{(1)}}+\frac{1}{b_{(1)}^2\sqrt{\lambda_2 n}}\left\{\lambda_1 b_{(1)}-\frac{1}{b}\left(\frac{\partial^2 b}{\partial\theta\partial\gamma}\right)\right\}\hat{U}-\frac{b_{(2)}}{b_{(1)}^2 n}\left(T_{(1)}+\frac{2}{b_{(1)}}\right)\\
&\quad+\frac{(\lambda_2-\lambda_1^2)}{2b_{(1)}\lambda_2 n}\hat{U}^2-\frac{1}{b_{(1)}^2\lambda_2 n}\left\{\frac{1}{2b}\left(\frac{\partial^3 b}{\partial\theta^2\partial\gamma}\right)-\frac{\lambda_1}{b}\left(\frac{\partial^2 b}{\partial\theta\partial\gamma}\right)\right\}\hat{U}^2\\
&\quad+\frac{1}{b_{(1)}^3\lambda_2 n}\left\{\lambda_1 b_{(1)}-\frac{1}{b}\left(\frac{\partial^2 b}{\partial\theta\partial\gamma}\right)\right\}^2\hat{U}^2+\frac{\pi_{(1)}}{b_{(1)}^2 n}+O_p\left(\frac{1}{n\sqrt{n}}\right).
\end{aligned} \tag{6.39}$$

Here, we have from (1.6), (2.1), and (6.3)

$$\frac{\partial^2 b}{\partial\theta\partial\gamma}=u(\gamma)bb_{(1)},\quad \frac{\partial^2 b}{\partial\theta^2}=b(\lambda_2+\lambda_1^2), \tag{6.40}$$

which imply

$$\frac{1}{b}\frac{\partial^3 b}{\partial\theta^2\partial\gamma}=\frac{\partial\lambda_2}{\partial\gamma}+2\lambda_1 u(\gamma)b_{(1)}+b_{(1)}(\lambda_2-\lambda_1^2). \tag{6.41}$$

Substituting (6.40) and (6.41) into (6.39), we have from (6.5)

$$\begin{aligned}
T_{B,\hat{\theta}_{ML}}&=n(\hat{\gamma}_{B,\hat{\theta}_{ML}}-\gamma)\\
&=T_{(1)}-\frac{1}{k}+\frac{\xi}{k\sqrt{\lambda_2 n}}\hat{U}-\frac{b_{(2)}}{k^2 n}\left(T_{(1)}-\frac{2}{k}\right)-\frac{\xi^2}{k\lambda_2 n}\hat{U}^2-\frac{1}{2k^2\lambda_2 n}\left(\frac{\partial\lambda_2}{\partial\gamma}\right)\hat{U}^2\\
&\quad+\frac{\pi_{(1)}}{k^2 n}+O_p\left(\frac{1}{n\sqrt{n}}\right),
\end{aligned} \tag{6.42}$$

where $\xi=u(\gamma)-\lambda_1$. Since

$$\xi=\frac{\partial}{\partial\theta}\log k=\frac{1}{k}\left(\frac{\partial k}{\partial\theta}\right),$$

it follows from (6.5) and (6.42) that

$$\begin{aligned} T_{B,\hat{\theta}_{ML}} &= T_{(1)} - \frac{1}{k} + \frac{1}{k^2\sqrt{\lambda_2 n}}\left(\frac{\partial k}{\partial \theta}\right)\left\{\hat{U} + \frac{1}{\sqrt{\lambda_2 n}}\left(\frac{1}{k}\left(\frac{\partial \lambda_1}{\partial \gamma}\right) + \frac{\lambda_3}{2\lambda_2}\right)\right\} \\ &\quad + \frac{1}{kn}\left(\frac{\partial}{\partial \gamma}\log k\right)T_{(1)} + \frac{1}{2k^2\lambda_2 n}\left\{\frac{\partial^2 k}{\partial \theta^2} - \frac{2}{k}\left(\frac{\partial k}{\partial \theta}\right)^2\right\}(\hat{U}^2 - 1) \\ &\quad + \frac{B}{kn} + O_p\left(\frac{1}{n\sqrt{n}}\right), \end{aligned} \tag{6.43}$$

where

$$\begin{aligned} B = &-\frac{1}{k\lambda_2}\left(\frac{\partial k}{\partial \theta}\right)\left\{\frac{1}{k}\left(\frac{\partial \lambda_1}{\partial \gamma}\right) + \frac{\lambda_3}{2\lambda_2}\right\} + \frac{1}{2k\lambda_2}\left\{\frac{\partial^2 k}{\partial \theta^2} - \frac{2}{k}\left(\frac{\partial k}{\partial \theta}\right)^2\right\} \\ &- \frac{2}{k}\left(\frac{\partial}{\partial \gamma}\log k\right) + \frac{\pi_{(1)}}{k}. \end{aligned} \tag{6.44}$$

Since, by (6.5) and $\partial\lambda_1/\partial\gamma = \partial b_{(1)}/\partial\theta = -(\partial k/\partial\theta)$, it follows from (6.44) that

$$B = -\frac{1}{2\lambda_2}\left\{\frac{\lambda_3}{k\lambda_2}\left(\frac{\partial k}{\partial \theta}\right) - \frac{1}{k}\left(\frac{\partial^2 k}{\partial \theta^2}\right)\right\} - \frac{1}{k}\left\{2\left(\frac{\partial}{\partial \gamma}\log k\right) - \pi_{(1)}\right\},$$

which, together with (6.43), yields (6.11). From (6.11) and Theorem 4.4.1, we have

$$T_{B,\hat{\theta}_{ML}} = T_{(1)}^{**} + \frac{B}{kn} + O_p\left(\frac{1}{n\sqrt{n}}\right),$$

hence, by (4.13) and (4.14), we obtain (6.12) and (6.13). Thus, we complete the proof.

Reference

Akahira, M. (2016). Second order asymptotic variance of the Bayes estimator of a truncation parameter for a one-sided truncated exponential family of distributions. *Journal of the Japan Statistical Society, 46*, 81–98.

Index

M. Akahira, *Statistical Estimation for Truncated Exponential Families*,
JSS Research Series in Statistics, DOI 10.1007/978-981-10-5296-5

Printed in the United States
By Bookmasters